Friends, Gardenline listeners and those of you who plan a vegetable garden, this is a complete guide on how to grow one successfully whether you live in the country, have a home in the city, or live in an apartment or condo.

I read one book that said, *"Vegetable gardening can be fun, relaxing and profitable."* I quarrel with that last word—profitable, at least the first time you try. But, there is no doubt vegetable gardening can be fun and relaxing. To me, it's therapeutic. And, there is **no better tasting** vegetable in the entire world than one you personally grow.

There is something so special about being able to plant a seed or small plant and nurture it to maturity. One Houston theologian said, *"If you think there's no God, how do you explain one small black seed becoming a giant watermelon?"* I'm certain that's not the exact quote but there really is a feeling that something extra special is going on when you plant.

I hope you enjoy my second book. I told so much about flowers, trees, grass, bushes and shrubs in my first book, **YOUR FRONT YARD**. This book on vegetables will be all together different. But now is the time for you to start planning. Let me help you along the way to make it easier and more enjoyable for you.

John Burrow

VEGETABLE GARDENING

Spring & Fall

VEGETABLE GARDENING

Spring & Fall

by

JOHN BURROW

SWAN PUBLISHING
New York • California • Texas

Author: John Burrow
Editor and Publisher: Pete Billac
Layout Artist: Sharon Davis
Cover Design: Mark Fornataro
Cover Photograph: Gary Bankhead

Books by John Burrow:

YOUR FRONT YARD
VEGETABLE GARDENING

First Printing, August 1995
Copyright @ John Burrow and SWAN Publishing Company
Library of Congress Catalog #95-070123
ISBN # 0-943629-17-9

VEGETABLE GARDENING is available in quantity discounts. Please address
all inquiries to SWAN PUBLISHING COMPANY, 126 Live Oak, Suite 100,
Alvin, TX 77511 (713) 388-2547

Printed in the United States of America

Dedication

To my mother and daddy, Roy and Neoma.

Introduction

In this book, I plan to take you, step-by-step, on the basics of planting a vegetable garden. You'll learn soil preparation, choosing the correct area for growing your vegetables, about care and maintenance and about every vegetable that grows in your particular area of the country.

I've written this book using 13-point print for those of you (like me) whose eyes have grown weak from watching too much TV and to save you time in having to search for your glasses to get some fast information when you need it.

I'll tell you some of the "basic basics", like the type of hoe, rake or shovel to choose. I did that (including pictures and drawings) in my **YOUR FRONT YARD** book. You ought to have a copy of that anyway.

I'll also tell you what I know about the "new things", the sprays, methods of planting, soil additives and changes that are far better than the old. Scientists and agricultural experts have spent years and many dollars to test and perfect changes that make vegetables grow larger, faster and healthier. Some even taste better.

Also, I forewarn you, I might get a little corny now and then and throw in a few stories along the way but all will be about gardening; remember I'm a country boy and it's just part of the way I was brought up and I happen to believe that a little down-home humor makes reading anything a bit more fun.

There are some "rules-of-thumb" in vegetable gardening that haven't changed through the ages and with the tens of thousands of callers I get over the KTRH Gardenline

show. I think the universal mistake most first-time gardeners make is in growing a garden that is far too large. The
smart way to go about it is, **grow only what you want to
eat.** I'm going to tell you about (almost) every vegetable
you can grow in this moderate to warm climate and about
some of the scientific methods that are best.

If you live anywhere in the world, everything I'm
telling you about growing vegetables applies. The only
things that change are the *kinds* of vegetables because of
climate. Follow these instructions I'm sharing with you and
your chances are high that your vegetable-growing experience will be a rewarding one.

JOHN BURROW and his co-host, **BILL ZAK**, can be heard on **GARDENLINE** every weekday morning from 10 till noon and Saturday mornings from 8-11am on **KTRH,** 740 on your AM radio dial. If you'd like to ask a question or speak with John Burrow, simply call **526-4-740.** If you have a GTE mobile phone, **KTRH** will pay for the entire call. Just punch in **STAR-4-740.** Long distance, dial **713-630-5-740.**

Table of Contents

Sprays and Pests
Varmits
Vegetables to Harvest Month-by-Month

Sweet Basil
Chives
Dill
Horseradish
Mints
Rosemary
Sage
Tarragon
Thyme

Containers and Planter Boxes
What ARE chives?

QUESTIONS AND ANSWERS

Part 1

SETTING UP A GARDEN

SIZE

Let's see now, where do we start? I guess to get it off my mind, I'll start with a story that will lead into this entire chapter. It's a true story and it's an important one. It could mean the difference in having fun with a garden or becoming a slave to it.

I had a friend who, for his first vegetable garden, started off with many tomato plants. He didn't know what type of tomatoes they were, he just gathered them from everywhere. He had the space and a new tiller, so he went wild. He didn't read a book nor did he ask any advice, he just happened to like tomatoes and heard from somewhere that they were easy to grow.

He started with 6 tomato plants that he bought from a hardware store. Then he got 4 more at Wal*Mart. A neighbor gave him a few plants, and a friend of another neighbor gave him 8 or 10 more.

He got a vegetable catalog in the mail and ordered some white tomatoes in a seed packet. While visiting a lawn and garden center, he learned of cherry tomatoes and bought two packs of those, 8 each to the carton. He had the space, so he just dug holes and planted.

When he began his garden it was to be just tomatoes and was a 10x20' plot that seemed big enough. When he set his plants down, he needed a little more room. Then he wanted to plant other vegetables (nobody has just a tomato garden), and his garden grew larger, to 20x40'.

Then, when he reasoned he'd have to have rows to walk down in order to harvest what he planted (and not step on everything else while doing it) and the garden became larger. With his (apparently) run-away tiller and a 5-gallon gas can, with all those plants, with a two-week vacation coming up *exactly* at planting time, his garden grew to about 50x90'.

Yep, those of you fortunate enough (or is in unfortunate?) to own a tiller, be careful with those things. It's fun to dig up earth and throw grass in every direction and when a beginning gardener gets their hands on one, it's like a kid with a Corvette; they just have to push it to the limit.

My friend's fun, relaxing little vegetable garden suddenly became expensive! He had to upgrade his garden tools and add to them. He bought a few types of hoes, two rakes, three types of shovels, a two-gallon sprayer, a large wheelbarrow, fence posts, garden hose, soaker hose, wire, about $500 of fertilizer and sprays and was spending upwards of 5 hours per day and most weekends setting up this garden for almost a full month.

And, everything grew! His tomato harvest was unbelievable. He was harvesting tomatoes daily in that deep wheelbarrow. He had tomatoes of all sizes and description, from the cherry tomatoes to the giant Beef-steak. He had tomatoes on shelves in his garage, the food bin of his refrigerator, he cut up tomatoes and froze them for his deep freeze.

He had tomatoes in hampers, bags, sacks and buckets. He ate more tomatoes than he had eaten is his life. He gave them to friends, neighbors, the mailman, people in various stores and carried them in his car in plastic bags and gave them to strangers.

The end result was that his garden simply became no fun, and ended up being almost a full-time job. Each tomato he and his family personally **ate** must have cost him about three dollars . . . per **bite!** His primary goal to relax with his first vegetable garden disappeared into **work!**

I don't want any of you to make this mistake. Start with a small garden and finish with only a slightly larger garden.I'm telling this story because it's probably happened to you or certainly someone you know.

Enough on what not to do. I just wanted to drive a point home on the size because whether we want to or not, the size grows faster as we add types and amounts of vegetables. I want you to enjoy growing and eating what you grow. Start small and let the size increase next year when you know more of what you're doing.

After you determine the size garden you feel you can build and maintain while enjoying it, it's time to look at the location of your garden. Here are some tips to help you.

LOCATION

Find an area in your yard where your plants will get sunshine. Your plants will need at least 6-8 hours of sun per day in order to grow. Make certain you're not under a large shade tree because the fruit will not mature properly, if at all. And either be near water or run a pipe (or hose) to that area. Sunshine and water are the two things all vegetables need.

There is a difference between **direct sun, sunshine** and sun**light!** The sun has to shine directly on your plants for them to grow best. Some plants grow in sun**light** only and the hot sun, does more damage than good.

It gets hot out there when that sun is right overhead but there's no need for you to get out there then. If you have work to do, do it before the sun gets hot, and when it's on it's way down.

Let me tell you a little about direct sun in any garden; it's strong and kills off many plants. If you have a small garden, it's wise to think about covering these plants to protect them from the intense heat we get from sun during June, July and August. When the sun gets hot with temperatures in the high 90's and runs into the 100's, **everything** takes a beating!

If you have a square to plant in, say 10x10' or 20x20', try to run your rows **east to west** to take advantage of sunshine. If you have rows and they happen to run north and south, plant the taller plants on the **north** side to prevent them from shading the smaller plants.

For instance, stalks of okra grow tall so those should be on the **north** side along with the smaller growing tomato plants next, bell peppers next, and smaller height vegetables in the first row.

Make a chart on paper and plan as if you're directing a movie or maybe taking a photograph. Put the tall people (plants) in back and the shorter in front so you can see everybody. Makes sense, doesn't it?

When you savor the taste of a complete vegetable salad that you grew, you'll be grinning like a Cheshire cat with pride. I guarantee that!

> I always wondered where many of these phrases came from but I happen to already know this one about "grinning like a Cheshire cat". It was from Lewis Carroll's story (and movie) *Alice in Wonderland,* where they showed this cat grinning all the time. Oftentimes when the scene changed, they would fade slowly away and only the grin of the cat could be seen.

If you haven't tried planting a vegetable garden, now is the time to start. If you're reluctant or don't have a lot of time, begin with containers even if you live in an apartment or condo or in an area were it's impossible to dig and plant. It's not like an out-of-body experience but it's an experience you'll never forget.

I look at is as one of those things you have to do and won't realize the pleasure of it until you do it. If you don't like it, well, you will because if you didn't like gardening and plants or flowers, you wouldn't be listening to this my program or reading this book. See, I gotcha' on that one.

TOOLS

Now that you have the size of your garden all planned out on paper and you have the right location, it's time to look into what you can get to make the work you'll have to do easier.

If you live in a home and have a flower garden, chances are you can use most of the tools you already have. I had pictures and descriptions of the tools I'd recommend for planting flowers and bushes and trees in my book, **YOUR FRONT YARD.** Get a copy of it for your plants, trees, flowers, grass and shrubs and take a picture of it all and send it in to the station. You might be a *Yard Of The Month winner.*

Let me tell you about tools. If you have plants, flowers and grass in your yard, it's wise to get the very best tools. Trust the fact, you'll break *five* $4 shovels before you break *one* of those $12 kind with a fiberglass handle. Same with hoes, rakes, and whatever other tools you might have for your garden.

I'm going to tell you about these tools to give you an idea on what you need. The smart thing to is for you to go down to your local lawn and garden store and look at all the tools. They invent so much so often that while I'm writing this book, manufacturers are coming out with new ideas that will make gardening easier for you.

I think if you **see** and **hold** and **heft** these tools, you'll get the ones that are right for you. I would recommend that you buy a light, long-handle shovel for your garden. You probably already have one of those so we've started by saving money.

Other than the "ordinary" shovel we all see with a long handle and a pointed end, there are **Scoop** shovels, used for picking up sawdust, fertilizer, manure or gravel. These are the really wide shovels with a short handle. You won't need one of those right now.

There is another shovel with a square point (long handle) and a **Transplanting** spade (also called a plumber's **Sharpshooter)** but you won't be transplanting anything now and if so, your regular shovel can get down deep enough.

You probably already have a rake too, but maybe not the one you need. I recommend the standard iron-tined **Bow** rake (beaux if you're from Louisiana or the counter part of an arrow—not bow like the front of a ship). The other rake is called a **Level Head** rake that might be the one you want to get to be able to get under some vegetables. Look at both but for now, get only one.

The standard garden hoe has a long handle and a 6" blade. You've seem 'em for years and might already have one of those too. But that is the only hoe you really need.

When you get serious about growing vegetables, you might want to add what is called a **Warren** hoe to your tool arsenal. It has sort of a triangular blade with a sharp point for making furrows or cultivating between plants.

Then there is the **weeding** hoe, that two-part thing with a hoe on one side and a gizmo on the other that helps pull weeds.

For tinkering after your garden grows and also for helping plant the seedlings or smaller plants you buy in pots, you'll need this combination set.

You'll see these garden tools at all prices and, of course, choose what suits your pocket book. I'm fore-warning you, these hand-held tools for 97 cents just do not work! Let me backtrack a bit; they'll work, but not for long! The handle is usually the first to go. It either slips, breaks off or cracks and falls off. The spade part of your trowel will probably bend and break with but slight pressure and, well, it's a mess. Get the small garden tools for about 3 bucks each. They'll last.

If you're following my plan of action, you'll be near beginning your garden. There are a few more things you must do in preparation to having a successful vegetable garden. You have the size, the location, the proper tools, and now you need to . . .

TEST FOR DRAINAGE

When you choose your planting area that has available water, dig a hole maybe a foot to a foot and a half deep with your spade or shovel and fill the hole with water to determine the **type of drainage** you have.

If this water disappears in 20-30 minutes, you've got a good start. If it drains in 10-15 minutes, that's even better. If it drains faster than that, you have far too much sand and need to add humus (black rich stuff at the bottom of your compost pile, or some good topsoil) to help the soil retain water long enough for your plants to take root properly.

Now, if that hole you dug and filled with water

doesn't drain very fast, say there is **still** water in that hole after it's set overnight, you'll have to make a **raised bed** with topsoil and some sharp sand to make it drain properly. If it drains too fast, the plants get thirsty and die. If it drains too slow, you'll drown your roots and they'll rot. This is your first step in **soil preparation.**

The next thing it's wise to do is to test your soil. You've already tested for water drainage, now let's find out the type of soil you have. I know this sounds like a lot of effort and you think it's starting to sound like work, but if you want a healthy, productive garden, do in right. Testing the soil is as smart as it is to build a strong foundation for your house. It will save you time and money in the long run.

SOIL PH

All through this description of plants and planting I mention the **pH** of soil. What is this little and big letter that are together with numbers following?

The **pH** is a symbol for the degree of **acidity** or **alkalinity** of soil. The number following reads on a scale of 1 to 14. A pH of **7** is neutral. A pH **higher** than 7 is alkaline and a pH **lower** than 7 is acid. The ideal vegetable garden is slightly acid to neutral with a pH of from 6-7. (Potatoes will grow in a slightly higher **alkaline** soil—pH 7-8).

To counteract the acid soil (usually in areas of heavy rainfall and with a pH of 5.5) you can add ground limestone (I'd use the **Hydrated lime** which is quick acting and applied several weeks prior to planting) or **Dolomitic** lime which contains both calcium and magnesium. It comes in sacks and you can get it at almost any nursery or lawn and garden store.

Alkaline soil is common in areas with low rainfall, poor drainage, and natural limestone deposits. To reduce the alkalinity, fertilize with an acid-type fertilizer and a soil acidifier. (*Easy Gro* and *Green Light* brands do the job).

To test the pH of your own soil, there is a probe that cost somewhere between 15 and 20 dollars and it will last you a long, long time. Just stick it in the ground and read the meter. Do it well ahead of the time you plan to start your garden so you can make the necessary corrections— each year.

Now that you've found the right spot, let's make certain all the weeds and grass are gone. I'd use *Roundup* or *Finale* to get all the trash grass and weeds out then till it and till it again. They are both good; (Finale works faster) but read the labels and see which is best for your particular garden area.

SOIL PREPARATION

It can take you about 3 or 4 weeks to prepare the soil for where you're going to plant your garden. In preparing your soil, if you have a tiller, use it. In fact, use it two or three times. Get the area where you plan to plant free of all rocks, grass, plant roots and other debris.

Pulling grass and weeds from a garden is the least

fun thing to do in the world, but it's necessary, because grass and weeds will take many of the nutrients and water from your vegetables. Use your rake, get a few of the neighborhood kids to help and/or prepare to do a little bending. Make that area plain and let's see only dirt (I mean soil). Dirt is what get under your fingernails whereas soil is what we plant in, right?

Use *Roundup* or *Finale* or any grass and weed killer your nursery guarantees will work. Trust me! An hour of two of pulling and raking will make you rush to your store for these products.

> If you're starting with a small garden, say 10x20'the smart way is to add additional top soil, sharp sand and organic matter and make a raised bed. This would be up about a foot above the ground. It drains better and your plants will respond better.

FERTILIZER

In the past, the only way to fertilize a garden was with animal manure. It is still an excellent fertilizer with a few shortcomings. For me, I'd take full advantage of modern science and go with the chemical fertilizer. It's easy and it works!

I advertise **Soil Pro** on my program over and over again and I do so not only because they pay for the ad, but because it works. When you start off your soil with a plus, like this soil additive and water, you give your plants a head start. This **Soil Pro** enhances the tilth of your soil, it retains up to 75% of its weight in water and nutrients, and releases them to the plant roots as the plants need them.

The most widely used manure in gardens is from a cow or steer. It provides nutrients to your plants as well as the addition of humus and desirable bacteria to the soil. For a vegetable garden, cow and/or steer manure will not give you maximum production and should not be used because it is not rich in food elements but rather, is a valuable soil **amendment.** How I'd use it is by allowing it to rot in a compost heap during the winter to condition the soil.

If your mind is set on going with the manure (let's say you have a friend who has a ranch who will give you the manure), here's what to do if you have say a 10x10' bed or 5, 2x10' rows, mix in 100 or so pounds of sheep or steer manure. Water the area after you've tilled or raked or hoed it in. It is just too hot to put in your garden raw.

Wait a few weeks and take the grass out that will grow up and till it again and two weeks later (one month total) do that again and you should be ready to plant.

My **YOUR FRONT YARD** book goes into detail about the nutrients in soil; *nitrogen, phosphorus and potassium.* But this is a different ball game with vegetables.

There are two general types of fertilizer, **organic** and **chemical**. I think the very best fertilizers are **chemical, slow-release** fertilizers (I've said that a time or two before) manufactured through chemical reactions in factories. These "people" know what's best for your garden or they wouldn't stay in business. And, chemical fertilizer works *faster* than organic.

With the **chemical** fertilizers, just read the label and follow the instructions. The only problem I can foresee (if everybody used chemical fertilizer) is on the environment. I do not, however, believe that everybody who lives in the city will use an over-amount of toxic, chemical fertilizer. READ THE LABEL!

Let me tell you why you need to fertilize. You might say, *"Plants and vegetables grew in forest before people were around to work it, weed it or water it"*.

That's true. But the problem with that theory is that people weren't around to eat the vegetables so they usually (the vegetables that weren't eaten by animals) fell to the ground and provided their own fertilizer. And the animals, when they came to eat, also *pooped*, which provide even more fertilizer!

Vegetables that are properly fertilized have several advantages:

1. It **doubles or triples** the yield because the plant grows larger.

2. The plant is **healthier,** thereby it **resists** the chance of harmful insects and disease. It also resists the strong sun that damages or kills unhealthy plants.

3. It **lives** longer.

On bags of fertilizer, you'll see numbers such as **10-10-10** or **10-5-15**. These numbers represent the amount of *nitrogen, phosphorous and potassium* in that bag of fertilizer. Sometimes you'll see the letters **NPK**. The **N** is for *nitrogen*, the **P** is for *potassium*, and the **K** is for *potassium*.

I know, why not **NPP**? Well, it's just that way. The **K** is the chemical symbol for *kalium* (also known as potassium). To further complicate matters, the last letter, **K** could also mean *potash.* Potash is the **oxide** of potassium, known as **Potassium Hydroxide.**

NITROGEN is essential for the green growth in plants. When your vegetables turn yellow instead of staying a nice, healthy green, choose to fertilize with a higher number first (the amount of *nitrogen* in that fertilizer).

PHOSPHOROUS is a nutrient that stimulates leaf and root growth. It will help the flavor of your fruits and will harden your plants for cold weather.

POTASSIUM or POTASH (KALIUM) also helps plants resist disease and damage from insects. Plants use potassium to maintain their salt balance. Potassium also promotes production of sugar, starches, and oil for food.

I like the **12-24-12** fertilizer before you plant. Then, to help while they are growing, the **12-24-12** mixed in water (one tablespoon full to a gallon) sloshed around the base of the plant. Makes a great starter solution.

If you read my first book you'll know that I grew up on a farm in Tulia, Texas, a small farming town about fifty miles south of Amarillo. My daddy had 160 acres of cotton and maize on some land he rented. He also raised chickens.

When I was a young teenager, I remember about fertilizer. My daddy had 5000 caged "layers" and I had to work with them. I fed them, watered them, picked up the eggs, "candled" them, sorted them, weighed them, then boxed and refrigerated them. Oh yes, the fertilizer. Can you imagine how much fertilizer is produced from 5000 caged chickens?

I decided not to get into the different types of animal manure because it just takes up space and each is best when aged—when added to a compost bin for at least six months. It's valuable when composted. Use the chemical stuff! I'm telling you, it's best.

COMPOST

This is such a major part in preparation of your vegetable garden and in growing healthy vegetables, that I want to spend some time with you on it and make it as easy to understand (for new gardeners) as possible.

Compost is the term applied to organic matter (leaves, weeds, grass clippings, sawdust, wood shavings, pine needles and garbage) which has been sufficiently decayed to form a light, crumbly mold. To grow healthy vegetables, compost is a necessity.

Let's start out by actually "building" a bin. If you have the space, the easiest way is to dig a hole in one section of your yard about five foot square and a 2' deep. In dry climates, maybe a foot deep. If the ground is extremely hard make it level with the ground.

If you live in the city and digging a hole is a problem, try making compost in plastic bags. Mix organic materials with fertilizers (one quart per bushel). The mixture should be kept slightly damp at all times. Then seal and store the bags at 70°F or higher. Don't use lime until the compost is ready to use in 3-6 months.

If there's such as thing as a *standard* compost heap and you have the room, I'd say the size should be about 4 feet wide, maybe 6 feet long, and 4 feet high, depending on the size of your garden. My publisher, months ago, (not good with tools) has a few acres to play with. After editing this book, he decided to have a vegetable garden of his own and proceeded to build a compost pile.

He started with one that is 8x8', 4' high because he didn't want to cut boards. He made a frame of 1x4 boards and made 4 individual "gates" using ordinary chicken wire and put latches on the top and bottom of each of these "gates" to form a box. Unbelievably, it's perfect.

Then, he started his "*lasagna*" *building* process by laying in a few inches of leaves, grass clippings and garbage. He threw in a little soil and maybe a few handfuls of 12-24-12 and sprinkled it with the water hose mixed with *Medina* (not too wet to where it's soggy). He then put down another layer, then another, wetting each layer until the pile "of stuff" was at the top of his new bin, and sat back to let time do the work.

If you live in the city and there's an odor, try using **superphosphate** (get it at any full-service nursery) on alternate layers.

Within a month, everything started to deteriorate. He raked it, used a shovel to stir it around, then bought a **pruning fork.** He used *Medina* to help promote bacteria and the end result, about 80 or so days after he began, was Humus—that black, soil that is truly wonderful for your vegetables.

In building a garden, the compost pile is really great to have. Each year you empty that compost pile, add it to the soil in your garden. I'd add a foot or two of humus in each row and till it all together and when the season comes around, plant. Do it year after year after year.

Okay now, let's backtrack a bit and start building your compost pile from scratch. First, make this "layer" maybe a foot deep and I'd start by throwing in some organic matter. You can use almost all the vegetables and fruit from your refrigerator that are beginning to spoil (actually any vegetable matter that will decay). Throw in some partly rotted leaves and grass clippings. I'd shy away from using a lot of leaves from a pecan or hickory tree; they contain chemicals that act like a herbicide.

Over this, lay 2-3" of soil, then the same amount of manure (cow, steer preferably). If you don't have or want to use manure, put in a 3" layer of peat moss. Then, sprinkle **raw ground limestone** over every other layer at a rate of a quart to a wheelbarrow load of compost. On alternate layers, pour about a quart of a complete chemical fertilizer such a *Easy Gro or Miracle Grow.*

If you choose not to build a bin and want only a "pile" just dump in weeds and sprinkle in garbage and throw in a layer of soil, manure or peat moss every 12-18" and soak with a hose. Add another layer, then another. It won't be pretty but it's fast and it'll work.

Never let the compost heap remain dry for any length of time; it needs to be wet to begin decomposing. At the end of every month, turn the entire heap inside out to accelerate decomposition.

The normal time it takes to make a successful compost heap is from 4 months to a year, depending on the rate of decay. If you do as I say and add the ingredients I mentioned and leaves and grass clippings that are partially decayed, probably 3-4 months. To maintain a constant supply of compost, you should start a new pile every six months.

Lime as well as *complete fertilizers* are also used for this "layer-building". Lime helps the pile break down (decompose) quickly. I'd not use the lime on the layer of fertilizer because together they cause a loss of nitrogen. Let each do their respective jobs.

Do not use coffee grounds. Tests at the USDA show them to be slightly toxic to plants. Same with sunflower seed hulls; they inhibit plant growth. Too, if you burn leaves and grass clippings, the ashes are of no real value. Don't let anyone tell you otherwise.

I cannot emphasize **soil preparation** enough. If you want a healthy, productive garden, you simply must prepare the soil. If you half-prepare, chance are high that you'll have a sickly garden. And, I don't care if the person next door has a terrific garden and does little or none of what I've advised. Don't take any chances and do it right!

Now that preparation has taken up a good part of this book, let's start thinking of . . .

WHAT TO PLANT

The smartest advice I can give you on this is to take a poll of those who live with you and try to plant what they (and you) like to eat. And start small because a few plants of certain vegetables is enough for a family of 10.

You're all set to begin now. You've got the list from your family members on the vegetables they prefer, now let's see if we can plant what they like in the area in which you live. Talk it over using this book as a guide to what you can plant, write it down and go out and buy them.

Be smart about this plant selection. Choose the plants that are easiest to grow and maintain, ones that are not too exotic, ones that grow fast, and bear the most. For instance, some tomatoes take 90 days and some take about 65 days to bear. That's almost a solid months difference.

Part 2

DIFFERENT GARDENS

CITY GARDENS

If you live in the city and have a small yard, the area you choose must have one main ingredient—sunshine. Remember, 6 to about 8 hours of sunshine per day is a must. I know it might be more comfortable to be able to sit under a large shade tree when you're planting or digging but the stuff just won't produce fruit (or vegetables).

If you have an area where the sun shines most of the day, say against your west fence line, that's perfect. If it happens to be in the center of your yard, that's just as good. If you're pressed for room the way a friend of mine was one time, build a **tier** garden like he did. Let me tell you how easy he said it was.

The first thing I did, John, was to get some **treated** *2x4's, 12' long, just four of them. I nailed them together forming a square and set them in an area that I sprayed with Roundup (they didn't have Finale then). When the grass and weeds were killed, I dug about a foot or more deep with a good shovel.*

I then got some steer manure from a friend who raised cattle just outside the city limits and dumped about 10 half-filled bushel bags of that into the center and raked it into the soil (didn't have a tiller then either). I also bought a few yards of bank sand I hauled myself with a small trailer.

I ended up with too much dirt and manure. So, like almost everybody else, I wanted to make my garden bigger. Instead of taking more land, I bought 4, 8' treated 2x4's, nailed them into a square, and laid them on the 2" over-the-ground first tier and now had framing for a **second** *deck.*

After I filled that in with bank sand and manure, I decided to really get fancy, so I bought 2 more 2x4's, cut those in half, nailed them together, and ended with a **third** *tier. I was able to utilize the extra dirt and manure and had a really great garden.*

If I recall, I had 5 tomato plants on the bottom level on the west side to let the sun reach my vegetables of a lesser height. I planted 4 bell pepper plants on the south side of that first row, I put 10 plants of Butter Crunch lettuce along the north side and the entire front, and on the east side, I went crazy with half green onions and half radishes.

On the second west side of the second tier, (4x4) I planted carrots on the south side, potatoes and beets on the north side, and on the east side I planted cabbage. On the third tier I planted three chili pepper plants on the back ⅓ and the rest in parsley. I had a really wonderful vegetable garden and that was all on a 8x8' piece of land.

It was easy to maintain, cost less than $200 to build including bank sand, free manure, lumber and the plants. (I used seed for carrots, beets and parsley, the poles I bought for the tomato plants, one quart bottle of Malathion and even some packets of Miracle Gro tomato food).

My garden took maybe 5 minutes a day to water. I got a soaker hose, sprung for a 9 dollar timer, and put it on my garden hose that I turned on for about 15 minutes on the warm days and maybe 25 on the hot days and the only real trouble I had was in wanting to spend more time on it but I didn't have to.

My friend swears that these figures are correct. If space is a problem and you want a backyard garden, you can do it without messing up much area. You can use that same space for a winter garden and you really will save money if you choose the plants I list in this book, plants that the entire household enjoys, of course. And, it really can be fun, not take a lot of time once the building and soil is ready, and a nice conversation piece for visiting friends.

> He said he even had enough extra vegetables to give to the friend who gave him the free steer manure.

Another idea you might try if you're short on space is

if you run your garden along and against your entire length of backyard fence (on the fence side to the west so it gets that sufficient sun). Make it about 4' deep so you can plant most of the "big stuff" in the center.

If you have a chain link fence, you won't need any wires, poles or cages for your tomatoes, just ties. Of course, I'd ask my neighbor if it was okay with them and perhaps bribe them with fresh tomatoes. Why not? They'll be able to take what they want anyway, especially if it grows on their side.

Should you have the standard 6' high wooden fence, you can use small staples and ties for your climbing tomato plants and also plant corn or okra along the fence and whatever other smaller sized vegetables you choose. Or, in front, if you make the garden say 6' deep, you will have sufficient room for a second row of something or other.

Then again, if you came out just a bit more, you could have a, uh, ah . . . a **divorce**, because now you're getting out of control and your spouse might not like that. Whatever you do, do keep your garden small enough so you can tinker with it and enjoy working in it.

You really don't need a large garden for a family of two, three or four. If you choose the right location, prepare the soil, water it, maintain it and grow the plants I recommend, it will surprise you how much food you will get for a small garden.

Let's suppose you have only a small patio, some with grass and mostly concrete. Then, you can build up a garden by using some of these treated fenceposts you see for sale at about $2.75 each. You can nail them as they are or cut grooves in each send so they will fit into each other and use long wood screws to secure them. They come in 8' lengths and you just have to cut them to size. Then, add your soil and make them 8, 10 or 12" high or higher and you have a bottomless planter box.

Many people who have decent soil for growing vegetables near the top of their ground and heavy clay underneath build these raised planter boxes and grow wonderful gardens. I would, however, use Finale or Round-up on the floor of the garden the till the soil or shovel it up to loosen it and then start building.

If you want to use 2x6's or 2x8's, get the treated lumber (tell the lumber dealer what you're using them for) and try this type of raised bed. Some people build it as high as 18" and some, who have concrete as the floor, go up as high as 24". This is how it's done.

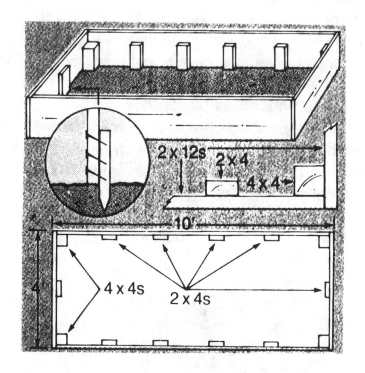

It's easy to build a raised bed.

SMALL GARDENS

These are the vegetables that can be grown in the least amount of space for those of you with no land (or maybe a border along one side of your cement patio). These vegetables can be grown in pots or planter boxes. They make interesting conversation pieces and come in handy for cooking.

Beets   Leeks
Broccoli   Mustard
Brussels Sprouts   Onions
Bush Beans   Peppers
Bush Squash   (Bell, Chili, Jalapeno)
Carrots   Radishes
Cauliflower   Spinach
Chard   Tomatoes
Chinese Cabbage   Turnips
Cress

I'm certain there are more vegetables to grow (maybe the sophisticated kind) but this is a good start. The smart thing to do is when you go to a nursery to look at these various plants, **ask the nurseryman** about what does what and how it suits your particular situation.

That's about it for City Gardening if you live in a house with a backyard. Now, follow the instructions that are universal for planting vegetables and you'll be okay.

The intense heat we get during June, July and August, with temperatures in the high 90's and into the 100's, causes everything to take a beating and can kill off many plants. If you have a small garden, it's wise to think about the location—before planting.

COUNTRY GARDENS

The same rules apply as with a city garden, the only difference is that you have more space and are prone to making it much bigger than you need. But if you have the room, the time it will take, and are willing to work, I say go for it!

If you have the space and the time, you could plant an acre! You could have beans and corn, watermelon and cantaloupe, onions, bell peppers, potatoes, cabbage, turnip and mustard greens; why not have it all! Maybe even have a 100 or more tomato plants, all kinds and all sizes. Go for the *Sweet 100* cherry tomatoes and try those 2-pound *Beefsteak.*

If you plan on a large garden, say 75x100' or more, you'll need a good **rotary tiller.** I recommend something that is affordable and before you buy, rent or borrow one to see if you can handle it. (*Ariens* makes a good tiller and so does *Toro*).

Shop around for the size tiller you feel suits you best and with the brand, (like anything else) make certain the factory gives you a good warranty and that they have service centers nearby.

Again, everything else you do with soil preparation, testing for drainage and fertilizer is the same as a city garden only on a much larger scale (depending on the size of your garden).

APARTMENT GARDENS

So, you have no backyard, maybe just a small patio that is concrete or perhaps just a balcony. You can still grow a variety of vegetables to eat and enjoy by using pots. In fact, this is much easier (not as many plants or varieties) but you need only the container(s), the proper soil, and you plant.

You will get fruit and vegetables and you don't have to weed or worry about testing your soil. You will, however, still need to spray for insects and diseases and to water regularly—not to much and not too little. And, if there is an unplanned freeze (this happens now and then) or a steamy day where everything outside bakes, you can cart the plants inside and protect them. I've known of many apartment dwellers who have year-round vegetables to eat that they grow on their own 6x10' patio or balcony.

You have a choice of containers to use. If you have a fenced patio, I saw some great old wine barrels cut in two at Wal* Mart a few days ago for less than $17 each that wouldn't look half bad on a patio with vegetables growing in them. Hey! Everybody has flowers on their patio but few have vegetable plants. It's wise, will save you money and is still decorative. Some vegetable plants are beautiful.

Of course, much of this "loveliness" doesn't stay long because you **eat it!** But it is a conversation piece and could make you different, in a "cool" way. (Do people still use the word "cool"?)

In the Houston area, for instance, the **Spring planting season** is usually around the middle to last of February, (hopefully) after the last freeze. Depending on the
temperature of the weather and the age of your plants
(other than seeds or seedlings) people plant as late as May
and early June.

The **Fall season** starts in September is perfect for
tomatoes. Again, dates mean little, it depends largely on
the climate. If you have a greenhouse or live in an apartment you can plant all year long and you determine the
season.

Further on in this book I tell you about the vegetables
that can be grown in this climate. I also have a special
section on herbs that apartment (condo, townhouse or
high-rise) dwellers might select from.

Part 3

VEGETABLES A TO Z

All of what I'm going to tell you on what to plant is true for city or country gardens as well as for those of you who live in an apartment, high-rise or condo, the only difference being the size of your garden.

This is the fun part but be careful now, you can't plant **everything** everyone likes or you'll need an acre or two of land. Remember, start with a few plants of a few vegetables and take it from there.

Below is a list of the vegetables that grow best in Harris County. I've mixed in Spring and Fall vegetables together to make it easier for you to determine what to plant and when.

To do this alphabetically I start with *Asparagus* and end with *Zucchini,* simply because I wanted to go from A to Z. I left out artichoke because they like cold weather, from about 0° to maybe just under freezing. It is a delicious vegetable (to me) in melted butter with a tinge of garlic. And those artichoke hearts are a true delicacy.

Too, even if we had this weather (which we don't) a single artichoke plant needs to be about 6' apart in rows that are from 8-10' apart. They simply take up too much room for the beginning gardener.

Some of you might like to know a little about *Avocados*. Well, for your information, avocado is a fruit (still delicious in salads) and it's "other" name is *Alligator Pear.*

I end my vegetable chapter with *Zucchini*, whereas zucchini should be under squash but it's the only vegetable that I can think of that starts with a Z.

ASPARAGUS

It takes a very a patient person to plant asparagus. However, it will reward you year after year—perhaps for a lifetime—because it is a *perennial.* The fact that it takes a full three years to harvest discourages many (unless you get stalks). I'd recommend trying it in your second year of planting; I'd like you to "test yourself" with your garden before you waste your time with such a slow-producing vegetable. Too, you might not even like asparagus.

If you do decide to plant it, asparagus loves a fertile, well-drained soil that runs from sandy loam to clay loam with a pH of 6.0 to about 6.7. Since asparagus loves cold winters best, the climate in this part of the South is not conducive to growing it.

> People have grown avocados, artichokes and asparagus in this area but few have success. Remember, I'm trying to get you instant success and enjoyment. Stick with the easy stuff at first.

BEANS ··

There are beans, beans, beans of every description. For instance, people want to know the difference between **string beans, stringless beans and snap beans**. This is the answer.

All older varieties of beans had strong, fibrous, **stringy** growths running the length of the pods. The removal of these strings was a tedious job. Then, plant breeders began producing varieties where the **strings** were eliminated. Thus, **stringless** beans. These stringless beans are easy to break or "snap" into pieces; hence they are now called **snap beans.**

For **BUSH BEANS**: *Tender Crop, Strike, Top Crop, Greencrop, Blue Lake, Jumbo and Contender.* **PINTO BEANS:** *Pinto 111, Luna.* **POLE BEANS:** *Stringless, Blue Lake, Kentucky Wonder, Dade, Romano.* **LIMA BEANS** (Bush type): *Jackson Wonder, Henderson, Fordhook 242.*

For the **Bush type** Lima Beans, they like well-maintained garden soil with a slightly acid pH factor. It's wise to plant them after the danger of frost has past. Seed at the rate of 3-4 per foot, 1-2" deep, with rows about 3' apart.

For pole type **LIMA BEANS:** *Florida Butter, Sieva, Carolina.* Pole type limas yield over a 4-5 week period depending on climatic conditions. They like it cool and not hot. Harvest them when they are full grown but still green. If they begin to turn white, they have been on the vine to long and become hard.

The pole limas take a long time to mature, some-where between 3-4 months so plant them immediately after the last frost is gone so you can eat them before the hot summer sun burns them.

For the best **RED KIDNEY BEANS in the world** buy the *BLUE RUNNER NEW ORLEANS STYLE RED BEANS* the next time you're anywhere in Louisiana. If you can't find 'em at the first store you go to, ask or try a second or third. The better restaurants in the entire state use these canned beans. They are delicious beyond belief!

The season for most beans is Spring or Fall and days to maturity range from 45-60 days. They are relatively easy to grow and you space the plants 3-4" apart. I like to space them 6" apart because crowding too much simply stymies the productivity. More isn't always better.

When planting beans, I'd space them 2-3' between rows to both let them grow freely and for you to be able to pick them when the do grow.

Start early with beans for **Fall** gardens. Make certain

it's tilled at least a foot or more and that you've prepared the soil correctly with the additives I mentioned earlier.

BEETS

The three main types for the Houston area are *Detroit Red, Green Top Bunching,* and *Pacemaker*.

Again, I like to start with plants and if so, space them 3-4" apart and a foot to two feet between rows, in soil that is fertile, well-drained and with a pH of somewhere around 6.5. Far too many first-time vegetable growers want to bunch things up. No, space them according to how I'm telling you. There will be enough for everybody.

Beets also are for Spring and Fall planting but prefer the cool weather. Like baby bear's porridge, not too hot and not too cold . . . just right.

BROCCOLI

There's *Green Comet, Emperor and Premium Crop* varieties. Most books say Spring or Fall planting but I'd say do it in the Fall because broccoli prefers cool, humid weather to the hot summers. Maturity time is 60-80 days. I've planted some in mid-September when the weather begins to cool off a bit and had broccoli all winter.

I'd recommend you plant broccoli 2' from each other or more because the plants sort of mushroom out and in rows 2-3' apart. Broccoli takes up a lot of space for the yield but in the Fall, you might have the room so, go for it!

The best soil for growing broccoli is fertile with a pH from 6.0 to 6.8. I'd set my plants 24-30" apart in rows that are 3 or more feet apart. If you are using seeds, plant them about an inch deep in the soil then thin them to that 24-30" spacing.

Keep soil moist (not soggy) and, of course, weed free. Water when it looks like it needs watering. If you use a *soaker hose*, I say turn it on 15 minutes every second day. Try not to water from the top because of the broccoli's penchant for leaf diseases

To **harvest** broccoli, use a sharp knife when flower heads form and while the florets (little flowers that make up one large flower as is the way broccoli looks) and cut a few inches below the head. If you cut too close to the head the plant will send out too many small side shoots rather than reform a new, large one.

If your broccoli starts to turn yellow and wilts, it's because it lacks water, fertilizer, or has a disease or root maggots. You can't cure it, dig it up and plant something else.

BRUSSELS SPROUTS ···

Two types, *Jade Cross and Catskill* are best for this region. They, like broccoli, are sensitive to heat and grow best in the Fall. It takes about 90 days for them to mature.

Plant them the same distance apart and in rows compara-
ble to broccoli.

Brussels sprouts like a rich, **friable** (easy to crumble
or pulverize) soil and need to be cultivated constantly. They
like water in dry weather and to be side dressed with
nitrogen or liquid manure during its early sages of growth.

To **harvest** these "baby cabbages" you simply cut
the "sprouts" off the stems with a sharp knife after the
leaves have broken off. Leaves usually snap off with ease
as far up the stem when the sprouts are ready to be cut.

CABBAGE

There is large variety of cabbage: *Early Round*
(sometimes called *Flat Dutch*), *Early Hersey Wakefield,
Savoy Hybrids—Golden Acres, Rio Verde,* the red *Ruby Ball
and Greenboy.* There is also a Chinese cabbage called
Michihli. Look on the label. I can't pronounce it but it's
spelled correctly. Then, plant your cabbage in rows that are
24-30" apart and no closer than 18-24" apart. It takes
cabbage (like brussels sprouts and broccoli) about 90 days
to mature.

Cabbage is a hardy plant and is easy to grow. Your
coleslaw will never taste as good when you grow your
own cabbage.

If your cabbage **heads** tend to crack, it's usually

because of rapid growth during warm weather on early cabbage. This causes a premature formation of seed stalk when the head is maturing and heads should be cut as soon as they are full grown.

To prevent worms from getting on your cabbage, dust the foliage with either *Dipel, or Sevin* when the worms make their appearance. The yellow and the white butterfly lay the cabbage worm eggs. For exercise, maybe get a net and try to catch them.

CANTALOUPE

I know its a **fruit** and not a vegetable but it grows well in this area, it will fit in many vegetable gardens, and it tastes so good, so I'm mentioning it. I can't figure why they don't spell it canta**LOPE** so everybody will know what we're talking about but I guess it's a good word for a Spelling Bee.

I love eating these melons, but growing them for most people (they tell me) is difficult. If space is a factor, it won't work; they need to be planted maybe 6-8' apart, the same as watermelon. But, some of my listeners grow them successfully and they plant them (if the weather isn't freezing) in early January and they mature sometimes in mid to late April before the weather gets too hot.

Most lawn and garden stores sell them in 4" plastic pots, but unless the weather is exactly "right" you just will not have any luck. I'd ask the local nursery owner or manager for some sort of guarantee or at least ask his advice. Maybe there are new species that I'm unaware of.

The vegetables we all eat, such as tomatoes and cantaloupe, never reach their full flavor in supermarkets. You see, these two are harvested **before** they reach their full maturity on the vine. They are picked **firm** (another name for half-green in order to be safely shipped by the wholesaler), then "color up" by the time you buy them but you get neither their full flavor or full food value.

CARROTS ··

I've never had any real luck growing carrots, they end up either too small or so crooked that it's unappetizing looking and I wind up at the grocery buying ones that are peeled and sliced.

But, many of my callers have had excellent luck with carrots in this Harris county area and they are best grown when the weather is cool. Plant them about 2-3" from each other and in rows that are at least two feet apart.

Carrots like deep, rich, loose soil and while writing this, I seem to recall that I probably never dug deep enough for my carrots to grow as intended. If the soil isn't broken up and loose and deep, the carrots take the path of least resistance and start to grow sideways and I end up with lumpy, stunted, crooked carrots.

> One thing I need to caution you about if you buy seeds of any kind; don't be afraid to pull them out for thinning. I know it's difficult to tear out a growing plant but most people seed carefully when they begin and end up scattering seeds everywhere. If these carrot seedlings are too close, **none** grow! Thin them out!

The type of carrots that grow best in this area are: *Imperator, Danvers 126, Nates, Red Core Chanatenay and Spartan Winner.* The fact of the matter is that almost any carrot sold in this area is good for this area, same as with any area. The experts are smart and their job is to sell merchandise and **successful-growing** merchandise will sell again next year. The send only the type of seeds, seedlings and plants that grow best in their respective areas.

CAULIFLOWER

I love cauliflower, raw in salads, to stick into a cheese or onion dip or boiled with butter. Don't fault me for my eating sins; this isn't a diet book.

Cauliflower likes cool weather and most types take 60-90 days to mature. Plant about 2' apart and in rows that are almost 3' apart because these vegetables grow wide.

The type I prefer to grow here in Texas is the *Snowball.* The *Snow Crown Hybrid* is also a local favorite and you can see from the word "snow" that it likes cold rather than cool, warm or hot. In hot weather, it rarely matures. Stick most of these plants at the top of your row so the water will drain off and cauliflower likes a fertile soil.

Cauliflower doesn't take up the room broccoli and brussels sprouts do but they need some room to grow so, set out plants about 2' apart. As they mature, I clip off the yellowed leaves and if they bolt, you'll know it because pretty little yellow flowers grow out (same with broccoli) and if you pick and eat these vegetables before the flowers get too tall, it's still good.

COLLARDS ···

You country folk will know what collards are and for those of you who don't, rather than have me tell you about them, go to your local supermarket and look at them in the bins.

The best type of collards to grow in this area are called *Georgia.* One book says Spring or Fall but collards prefer it cool, like cabbage, broccoli, brussels sprouts, et al in that family.

It takes collards about 75 days to mature and to plant them, I'd space them 2' apart in rows that are **at least** 2' apart. Give yourself room to weed, pull off the dead leaves and to harvest them.

Collards like rich soil and cold climates. I'd call them a **Winter** crop and plant them in September through January but they are hardy plants and can also take the summer heat. Try your luck with them and let me know how you fare. A popular name for collards is "headless cabbage" and a well-nourished plant can grow to 3' tall and yield many juicy and delicious leaves.

CORN ···

There's a **sweet** corn and a **white** corn. My choice for the white corn is *Silver Queen* and for the sweet type, either *Merit, Calumet, Florida Staysweet, Guardian, Butter-sweet or Funk's G90.*

There's a variety called *Early Sunglow* that grows in cool weather and the stalks reach a height of 4-5 feet. Plant your corn about 1' apart and in rows that are as much as 3' apart because corn needs room to grow.

There are several new types of **Supersweet** corn that are higher in sugar content and hold their sweetness longer. Ask your nurseryman to advise you on the kind that grow best in your area.

Corn is one of the vegetables that grow in Spring or Fall. The Spring planting dates are from the beginning of March until mid-April. The Fall planting dates are from August the 1st to about the 15th. Plan on it taking about 90 days to mature. And, I know you hear the phrase "rows of corn" but I'd suggest it be planted in **blocks**; actually **rows** of blocks.

Let's talk more about corn for a moment or two. This "row of corn" business is misleading because corn needs to be in rows but also **blocks** or rows so there's no need to try running a **row** or corn along a fence for the simple reason that corn must be pollinated.

The corn pollen from the tassels of one stalk is carried by the winds to the silks of the next stalk and corn planted in a block has the best chance of getting this necessary pollination.

And, to successfully grow corn, you'll need to satisfy it's three basic requirements; (1) Space, (2) Warm weather, (3) Generous amounts of fertilizer and water.

Corn needs at least 8 hours of maximum sunlight and has to be relatively by itself because it grows so tall that it shuts out the sunlight for other plants. If you plan to grow more than one variety, space them far enough apart so they can't **cross-pollinate**.

Enough of this "corn talk" for now, okay? At the beginning, I said "a minute or two" and there's so much to say that it would take maybe an hour or more. If you really want to grow corn, get either of the types I mentioned earlier and plant them in a row of blocks in the corner where they get full sunlight. Then, fertilize and water continuously and remember, plant a lot because one or two ears off of each stalk won't feed a family for very long.

When I was in probably the 8th grade or 9th, we had a garden with four acres of corn. I can remember when the corn was ready, and my daddy decided to sell it at the side of the road. People would drive up when we sold all we had, daddy would say, "Let's go cut some corn".

He'd pull the ears off and I was carrying a burlap sack for him to throw it in. I recall the ground was always wet because we wanted to keep the corn wet. I was a big boy then but carrying a full sack of roasting ears back to the road was not easy. It took a long time before I liked eating corn on the cob again after that "learning experience".

If you want to get some **fresh** corn at a deal price, drive Hwy 59 or 90 South towards El Campo and there are corn field that run as far as you can see. Usually, there are stands set up alongside the road.

CUCUMBERS ···

I know I should have said "cucumber" (singular) but nobody grows one cucumber. Here in Texas, especially in Harris county, you can choose from the *Victor, Liberty, National Pickling, the Ohio MR17, or the SMR58.* Spring is the best season for planting cucumber(s), anytime between March 15th and May 1st.

Since they need so much room to "roam" I'd recommend you plant them at least 2' apart. One book suggests rows any which way.

Cucumbers like rich soil that is well-drained. I grew mine on a slope and had excellent results. I'd water with a soaker hose and let the water run to the roots then run past them. Of course, I watered every morning usually before golf.

Cucumbers are great in salads. One of my favorite combinations is cucumber, bell pepper, celery, tomatoes and onion in an oil and vinegar dressing. In my garden, a cucumber would be one inch long one day and ready to pick the next. Pick them before they grow to about 6" long; the really big ones are tough.

I always recommend that you read the label or instructions on every packet, bottle or box of anything that you buy because nobody knows better on how this particular product should be used than the manufacturer of that product. If you're not certain, always ask your local nurseryman or call the Gardenline # and ask me. If we don't know the answer, we'll find it for you.

EGGPLANT ···

The eggplant is actually a fruit, commonly called a vegetable (same as tomato and bell pepper). The planting date for eggplant in this region is between March 15th and April 15th because it is a warm weather crop. I'd say to plant seedlings 1" into the earth, about 2' apart and in rows that are nearly 3' apart. Keep them moist and fertilized.

Eggplant wander along the earth but I've trained some to go up poles thereby needing less space to grow. Just be certain to pick them off when they ripen or they'll fall to the ground and maybe break.

I think eggplant add color to your vegetable garden,

all plump, purple and polished to an inky sheen. It takes maybe 80 or as many as 90 warm days for an eggplant to mature, sometimes as long as 100 days. If you have a short summer, choose the fast-maturing type *(Dusky)* that reach full maturity in as little as 50 days.

The yield of a single eggplant bush is maybe 6 per plant. Harvest them before they over-ripen or crack because of the intense sun. When clipping, use a knife or scissors so as not to tear the stalk.

There are a variety of eggplant, not all are the often seen identifiable purplish black. They are also yellow, green, and white as well as smaller types that come in more rounded or cylindrical shapes. The best kind of us to plant is: *Florida Market, Black Beauty and Ichiban, Florida High Bush and Dusky.*

GARLIC ···

The kind we most use is called the *Texas White*. It's a fall plant (about mid-September to maybe mid-October) and takes a long time to mature, from 5-6 months!

You can start off by buying bulbs from a nursery that are virtually pest-free. You can also buy garlic from a grocery store and many have done so with great luck. Garlic needs a sunny spot with rich, well-drained soil and when planting, use only the biggest cloves, better with pieces of root attached.

In planting, set out these good cloves in soil maybe

1" deep and about 6-8" apart. Plant cloves with **pointed** ends up. If you have the larger variety (*Elephant Garlic*) space them about a foot apart still pointed end up, an inch under the soil, and maybe a foot apart.

The soil must be kept moist, weed regularly and pinch off any blossoms that form. When leaf tips start to turn a yellowish-brown, stop watering and step on foliage, pressing it flat to the ground. This prevents further flowering and hastens maturation. Harvest bulbs when leaves are mostly brown.

Now, don't just dig down with a shovel or try to jerk it out with a hoe or try to pull the garlic up with your hands, use a garden fork and gently lift them up. With your hands, you might crack the bulbs and decrease the storage life of your garlic. It's wise to let the garlic bulbs dry outside in the sun (maybe 3 weeks) until the skin is papery and don't let it rain on them.

> It seems like a lot of trouble to me and, unless you love garlic, I'd go to the grocery and buy some as needed. I'm just mentioning this because you can grow garlic—if you like.

LETTUCE

There are five categories of lettuce. There's the **Butterhead** (sometimes known as *Bibb lettuce*), **Leaf** (also called *Loosehead lettuce*), **Romaine** (*Cos*), **Crisphead** (oftentimes called *Iceberg*), and **Celtuce,** which resembles a cross between celery and lettuce.

With the Butterhead, you have the *Summer Bibb, Tender Crisp*, and the tasty *Butter Crunch.* With the Leaf lettuce, you have the *Black Seeded, Simpson, Salad Bowl, Ruby*, and the *Oakleaf.* With the Romaine lettuce, you have *Valmaine*. I like them all!

If you opt to buy the packets of lettuce seed, the name and a picture of the lettuce is on that pack for you to see what you're buying. If you do choose the seed, after your bed is prepared, run a rake handle along the top of the row and form a small trench. Then, scatter the seed in a row and cover them.

When the seedlings appear, don't be afraid to toss the ones that are too close to each other away to make a 4-6" space between plants and about 18-24" between rows.

The Spring date for planting is from February 1st to April Fools day; the Fall date if from September 25th to about October 15th. Either time, it takes about 70 or so days to be ready to eat.

Lettuce favors cool weather so I don't especially agree with the date of the spring planting. Since it takes lettuce as short as 50 and as long as 90 days to mature, if it takes the full time that brings us into the warm months and the lettuce turns brown and bitter, so you decide. If you have a mild winter, I'd plant in mid-January or early in February. You always take the chance of a "freak" freeze setting in and killing it.

Another tip on planting lettuce where the weather

might be too warm is to do so in the shade of taller plants such as broccoli or corn, even tomato plants. Makes sense; lettuce and tomatoes.

Leaf lettuce is the easiest to grow because of its soil tolerance. It will do fine in poor soil with good drainage and light fertilization at planting time. Leaf lettuce comes in red, bronze, dark green and chartreuse and the leaves are exceptionally tender.

For apartment dwellers who grow vegetables in containers, leaf lettuce needs only about 12" of soil because of their shallow roots. Add cherry tomatoes and you can have all the trimmings for a hamburger cooked on your patio grill.

MUSTARD ·

I like the *Tender Green* (has a sort of spinach flavor) and *Florida Broadleaf* (has smooth, easy-to-wash leaves). Planting time in Spring is February 15th to April 1st and in Fall, September 1st to November 1st. The plants should be 6-10" apart in rows that are 2' apart. They mature in 40-50 days.

The Mustard, or Mustard **Greens**, have tender leaves somewhat like a watercress tang and are excellent for salads. The mustard greens favor cool weather and too much heat makes the leaves tough and the taste tart.

In well-moistened soil, greens spring up in about a

week after planting seeds (as little as ¼" deep) and when the plants reach a height of maybe 5", it's time to thin them out. Caring for them is easy, keep soil moist and keep weeds out.

When the lower leaves mature to the size of your outstretched hand, it's time to harvest. The upper leaves, not as tender, are also good especially for cooking.

OKRA ···

People in Louisiana love okra but the slippery texture of stewed okra takes some getting used to. But fried, sauteed or steamed, a crisper texture emerges. The Cajuns use okra for flavor and to thicken their soup and their world famous *gumbo.*

The word *gumbo* has two different meanings, depending whether you live in Louisiana or Texas. I'm not certain where the word originated. Perhaps its a cross between what we know as *gum* and the French word *rubbery.* Or maybe it's from the Angolan word *gombo* meaning *okra.*

I like the *Clemson Spineless, Louisiana Green Long Pod* that gets to maybe 2½' tall.

Again, you can use seeds and plant them maybe ½-1" deep in a row and 6" apart. Thin the seedlings to about 18"-2' apart in rows that are about 4' apart. These plants grow tall and strong. If you plant them too close together, you'll need a suit of mail to harvest them.

It takes about 60 or so days for okra to mature. Pick the pods are from 2-4" long. Any larger and the pod toughens. Okra appears almost overnight so keep a watchful eye and harvest it at **least** every two days. I had some okra one time and I went away on a weeks vacation. When I returned, the pods were over a foot long and dulled my crosscut saw in trying to get them off the bush.

I'd also suggest you wear garden gloves when working around okra unless you plant the Clemson Spine-less. And, you can plant the Dwarf Green variety in a large pot or tub on your patio.

ONIONS ·

I'll only tell you about the **Green** onions; the *Beltsville Bunching, Crystal Wax and South Port White*. The others (white, yellow and red) that you see in the supermarket, take up a lot of room and are inexpensive even during the highest-selling times.

I'd not start out with the seeds because you can buy a small bunch at almost any nursery for a dollar or two and stick them in the ground and they'll sprout. These are also excellent for growing in pots on your patio or back porch to just reach outside, clip off a few of the green tops and put in on your baked potato, sandwich or to flavor almost any dish.

To get your onions off to a rapid start, plant in cool weather; they favor cold climates. They like rich, well-drained soil with a pH of 6.0 to 6.8. A good potting soil (for containers) has everything you need in it.

In your garden, set them out in Spring about 4-6 weeks before the last frost date. My onions grow all year round. Plant them about 2" deep maybe 4" apart in rows spaced for 20-24" apart and apply liquid fertilizer. I've even planted in mid-summer but, in the shade where only sunlight hits them.

PEAS ··

Peas are fun to grow because you can see the results of your efforts quickly. Also, they taste good and (again) you have never tasted a pea until you've had one from your own garden. I'll tell you about only a few varieties of peas, ones that grow best in any area of the country.

First, there's the **English Pea,** by far the favorite among gardeners, of which 3 out of 5 grow them. Some are for shelling, some have edible pods and some, either way. The garden peas grow on a vine. Dwarf peas grow to maybe a foot-and-a-half to two feet tall. The tall varieties grow to 6 and even 8' high and need poles, or string or chicken wire fences to climb. The **Southern Pea** mature in from 50-70 days for varieties like the *Blackeyed Pea* (New Years Eve dish), *Crowder Pea,* or the *Purple-Hull Pea.*

Plant peas in rich soil with the pH between 5.5 to 6.8.

You must sow the seeds 1-2" deep, an inch apart, and in single rows spaced from 3-4' apart. You must thin the seedlings to about 4" apart and begin making plans to set up poles, wire fence or a trellis of some sort for them to grow. They yield from 2-6 pounds per 10' row.

Before planting seeds, soak them overnight in water. Make certain the soil where you plant is moist but seeds will rot if you water too frequently, at least until they are out of the ground and thinned out.

Harvest the pea pods when the pods are 2-3" long and when you can see that the peas in these pods are almost round shaped. For the freshest taste, harvest every 2 days.

All peas like well-drained soil that is rich in organic matter (compost, leaf mold, bone meal, and manure) but limited in nitrogen. Too much nitrogen will produce lovely foliage but few peas. And dig deep, make that trench (preferably) about 2' deep. Make it easy for those roots to form.

*My mother grew blackeyed peas she sold for some-where around 2 dollars (maybe 4 dollars) a bushel. I remember one year when I was about 12-years old there was a shortage of blackeyed peas and the price shot up to **twelve** dollars a bushel. We all went to work on the peas and thought we were rich.*

PEPPERS ···

There's the *Cayenne, Tobasco, Fresno Chile, Serrano Chile, Hungarian Wax,* and the all-time Texas favorite, the **Jalapeno!**

These peppers range from mild to searing hot and range in size from the 1-1½" long *Tabasco* to the 7" long *Pasilla* peppers and in colors of green, red and shades of brown.

I'd suggest, especially in this case, that you read the labels carefully! I grew up in Tulia, Texas and I can take some spicy foods and hot peppers but some of these peppers are truly lethal. Plant what you can eat but please, know what you plant.

Then, there's the favorite of many, the sweet **Bell Pepper,** so named because of it's bell-like shape. These Bell peppers are harvested before they mature, when they are green. Most people think they are supposed to be green but when they fully mature they turn either red or yellow, depending on the variety.

Best for most areas are the *Tambrel 2, Shamrock, Gypsy*, and the *Belltower*. There is also the *Yellow Banana Pepper,* the *Sweet Hungarian Pepper* that grows from 4-6" long and is slender and pointed. They are harvested when they are yellow but when they mature they turn red. Then there are *Cherry Peppers* that are globe shaped, an inch or two wide and also sweet.

Again, check the back of the packet for instructions on how to plant. If you see a pepper growing on a bush and you decide to taste it, do so carefully. If a mischievous friend hands you one to "toss down" don't do it.

Peppers of all varieties are somewhat demanding but well worth the demands. They like a sunny location, fertile soil, ample moisture, protection from strong winds, weeds taken from their root areas, warm days and slightly cooler nights. I've never had trouble growing Bell Peppers (or any other kind) in the south.

For Spring planting, don't start before the last freeze and for Fall planting, the last two weeks of July and expect maturity in 60-95 days. Set plants about 2' apart in rows that are 3' apart.

Harvest Jalapenos when they are dark green. Allow Pimentos to turn fully red on plants, the sweet peppers when they are full-sized; the hot peppers when they are full-sized and turn yellow or red; and the bell peppers, green, red, or yellow; just don't let them get too large.

POTATOES ···

I'm not in favor of planting potatoes because they are so cheap to buy, but I'll tell you a bit about the background and heritage of the potato. Unlike what most people were taught to believe, the **Irish Potato** did not come from Ireland; it was carried to Europe by Spanish explorers from South America.

Then, Irish farmers discovered their food value and

helped make the potato popular, so popular that the crop native to South America is known worldwide as the Irish Potato.

Irish settlers brought these potatoes to the colony of New Hampshire in the early 1700's and eventually farmers in the United States grew and developed many other varieties.

It was perfect for settlers who farmed to eat. It was easy to grow, long-lasting if stored properly, thrifty, nutritious, heartily satisfying, and wonderful to eat, especially with butter, sour cream, chives and bacon bits, huh? Most especially if it's accompanied by a large, juicy, sizzling steak.

If you do go against my advice and decide to grow potatoes, get **certified disease-free seed potatoes** from a nursery. They come in numerous varieties from feed and seed stores, farm supply stores and through mail order in red, brown or white (actually pale yellow) skins. Most varieties mature in about 3 months.

Plant in fertile soil that is well-drained (preferably sandy loam) with a pH of 4.8 to 5.4. Plant in cold winter climates or in Spring, 4-6 weeks before last frost date.

Cut these seed potatoes into blocky chunks, each with 2 eyes (growth buds) in furrows 4" deep, 12-18" apart, in rows spaced about 3' apart. Cover potatoes with about 2" of soil and add 2 more inches of soil when sprouts emerge.

Keep soil uniformly moist and weed regularly. "Mound up" soil around the vines as vine grow, to keep growing tubers protected from sunburn. In harvesting, dig beneath the plants with a spading fork or shovel and keep tool about 8-10" from plants to avoid injuring the potatoes. Then, lift the plant gently, shaking off the loose soil and pull potatoes from vines.

> At planting time with potatoes of all varieties, fertilize in bands along both sides of each furrow, keeping fertilizer about 2" away from seed potatoes but at the same soil level. Use a 10-10-10 fertilizer at about 8 pounds per 100' of row—half this amount if manure was previously tilled into the soil.

Harvest time is when most of the foliage has turned yellow to brown. Then, water plants for the last time, the wait an week to 10 days and cut away vines. This tends to set or harden the potato skins so they won't peel or bruise too easily.

In another 5-7 days, preferably when it's cool or overcast, dig up the potatoes (spading fork or shovel) and gather the harvest in burlap bags or baskets. To heal any injuries, store potatoes for 2 weeks in a dark place with high humidity and in 50-60°'s.

Keep only healed specimens for further storage and don't expose them to light for any length of time. Further storage should be in a well-ventilated, dark and dry location such as a basement where the temperature is around 40°F. If it's warmer, potatoes may sprout and if cooler, their starch may turn to sugar, sweetening the flavor. Potatoes should keep well for 3-6 months.

SEE . . . why I suggested you just buy potatoes at the store? You have to either be in the potato-selling business or really love potatoes to do all of this when a few bucks can buy a sack full. Also, in storage and before cooking, cut off and discard any green portions of the potato. **They are poisonous!**

RADISHES

These are hardy little fellas' in that they can stand cold winter but they also enjoy mild winter climate, when the temperature is about 40°F, of course.

The ones I like are the *Black Spanish* and *White Chinese* variety which can be found in many stores in packets. Since radishes come in all shapes and sizes (small bite-sized to as long as one foot long or even the size of a turnip or a cucumber), radishes are from mild to sizzling hot.

To grow, like most vegetables, they need a sunny spot and soil that is tilled 8-10" more that you expect the radishes to grow so as not to stunt their growth.

Radishes can be planted almost in any soil and they'll thrive. Children love them and they pop up in as little as three weeks on until about two and a half months. I run my rows along my East to West fenceline and let them do their own stuff. Some are small and some larger. The mistake I make is in not thinning them out enough and one grows into the other and I get small radishes.

For a garden the kids will enjoy, rake the fine ground soil into little ridges maybe 6" high and a few feet apart. The children delight when they see results in about a week after planting the seeds. For Spring planting, start February 1st to about April 15th and for Fall planting, September 1st to November 15th. And when you harvest one crop, dig up the soil, add some organic matter, and plant again. You'll have radishes all year long.

To care for them, water regularly and keep weeds out and don't forget, **thin** them out so they will have room to grow. It's better to have a few pounds of tasty radishes that a whole pile of leaves with nothing on the bottom. They yield from 2 to maybe 5 pounds per 10' row. You can **pull up** the small radishes but the larger ones need to be dug out.

Water and sun are the main ingredients for almost all vegetables. Tilling the soil, adding fertilizer, keeping weeds out and thinning your seedlings are all important. As I often say, "The best thing to put over your garden is your shadow."

SHALLOTS

They look like onions but have a more pungent flavor with a strong trace of garlic. These are the suckers I like on my baked potato and they are grown easily. They can take cold winter climates yet they thrive in mild winter when the temperature is around 40-50°F. For a Summer crop, plant about a month before the last frost date and in mild Winter climates, about 6-8 weeks before last frost date.

Shallots can be bought at most grocery stores or ordered by mail. If you have bulbs, plant them with the fat base down and just cover the tips in soil. They like a fertile soil with the pH from 5.0 to 6.8. What I do is make a section of my garden closest to the house in a square. I plant the bulbs ¼" deep about 6" apart and they start growing in about two months.

Then, I go outside as I need them, cut or pinch off the long green leaves and cut them up to add flavor to soups, salads, baked potatos, steaks, meat loaf—just about everything! When one section looks as if it's becoming vacant, I slip in a few bulbs and take my chances. I have shallots all year long.

For replanting, dig up the bulbs when the tops turn a yellowish-brown and seem to die. You can store them for as long as six months and they are hardy. You've all seen them being sold at nurseries, Wal*Mart or K-Mart where they look dead—but they aren't. Prepare your soil, get that pH factor where they like it and they'll grow.

I can't emphasize enough the importance of reading the label on whatever you purchase, whether it's fertilizer, an insecticide, or a packet or package of plants. If there isn't a label, ask the person who looks like they know what they're talking about. Information is everything!

SPINACH

Popeye might like it but kids, from my experience, do not! In the gardens of ancient Persia, spinach was the delight of royalty as well as it is with gourmets of today because it just seems to "go" with so many things. It's rich in vitamins and minerals, great in salads, lends color and spice to quiche and soups. I like to boil it and take a large serving, add a slab of butter, a sprinkle of salt, a longer sprinkle of pepper, and make an entire meal of it.

Choose from the *Early Hybrid 7* or *Melody in Fall* and try either the *New Zealand or Malabar* for a Summer crop. Spinach prefers cold and/or cool weather with temperatures around 60-65° but can hold up to near freezing weather.

If it's too long before harvest or if it gets too warm, it bolts and flowers and the flavor is gone. Better to dig it up and try again or to plant something else in its place. That's why I suggested the *New Zealand or Malabar* varieties because they are more accustomed to warm climates. (*Malabar* spinach can be trained on a trellis). Remember, look for the name and read the label. It will save you time and a migraine.

Spinach thrives in cold weather and in rich, well-drained soil with a pH factor from 6.0 to 6.8. Now, I know many of you beginning gardeners **bolt** over this pH business but it's necessary and you can find out what it is with an easy test and maneuver the type of fertilizer to that particular area. It pays off friends!

In fact, maybe now's the time to go into this pH thing and tell you how to do it and show you how easy it is. It really is the difference in plants flourishing or looking like dying grass. It's the difference in growing vegetables or deformed clumps of things that resemble produce.

SQUASH ···

Summer squash should be planted between March 10th and the last day to file a tax return, April 15th. The ideal soil is rich, well-drained, sand or clay loam and with a pH factor from 5.5 to 6.8.

If you're using seeds, plant them about 3" deep, 15 or so inches apart in rows spaced about 8' apart. When the seedlings appear, **thin them out,** from 4-6' apart because squash takes up lots of room. I like to really give my squash room to grow and I position the seedlings in groups of 4 or 5 as much as 8' apart in rows that are 12' away from the next one. I don't like to see the vines creeping and crawling and tangling with each other. Needless to say, squash is not the best plant for those with small areas to work with.

The varieties of squash are an important factor in that the **Summer** squash I like is *Zucchini* and *Dixie*. There is also a Summer squash called *Hyrific, Hybrid Crookneck, White Bush Scallop, and Multipik.* Oh, I like that *Butterbar* type too.

If space is a factor, shy away from the vines of Summer squash; they can be as large as 4-5' across at maturity. See why I give them plenty of room? Choose the **bush** type—short vines and still large plants. The soil should be rich, well-drained, sandy or clay loam and the pH from 5.5 to 6.8. The yield is bordering the unbelievable in that you get as much as 80 pounds of squash in a 10' row. Then again, you could get as little as only 10 pounds down to none.

The time to harvest depends on the growth factor. It takes about a full month for **Summer** squash and a month-and a half or so for **Winter** squash. What I suggest is to watch it's growth and when your Summer squash is tender enough for you to poke the skin with your fingernail, take it inside and cook it.

With the **Winter** squash, take it in when the skins are hard and not scratchable, usually in late autumn after the vines have dried but before the first heavy frost. When you determine that you can't scratch it with your thumbnail and when the stems are thick, cut the stems with a sharp knife or scissors leaving a 2" stub on the fruit.

Squash will grow on a compost pile because they love lots of organic matter. They need regular watering (they all do, don't they?) and good drainage. Having water surround and sit on a root is as bad as not having any water at all. And with the watering, overhead watering is not recommended; they like deep watering. Overhead watering is not recommend but for a few plants because it encourages leaf disease. Too, they seem to grow overnight and with reckless abandon.

I looked at a zucchini squash on a Monday morning just before I went to the station to do the Gardenline show. Tuesday I had the day off so I played golf. I again looked at what was the small zucchini squash on a Wednesday afternoon (after my program) and it was **10 inches long!**

The Buttercup squash that I like, grows in sort of "turbans" that are 4-5" long, 6-8" wide, and dark green, gray green, or orange in color. Squash can be prepared in a variety of ways using a variety of spices. And, either you like squash you don't. If you really don't like squash, I'll give you my mama's recipe and I promise you'll change your mind.

Another time I remember well was when I was discharged from the navy. My daddy had a truck farm consisting of 5 acres of squash and 5 acres of turnips. When it was squash-picking time and there was a rain, we'd get 240 bushels of squash every other day. If it was really wet, the squash got too big and you'd have only hog food. So wet or dry, I was sent in to cart it out. I'm not that fond of squash either.

Whatever corn we couldn't sell by the side of the road

still had to be "handled". Yep, I spent a lot of hours washing, picking and culling bushel baskets of corn only to haul off to a grocery store and hope they buy it when we got there. It's a tough life.

Let me try to give you an idea of how large a bushel is so you can understand how much work I had to do to haul off 240 bushels of squash. 1 pint is equal to almost one half of a liter. A **bushel** is equal to about **70** liters. It takes 8 **quarts** to equal 1 **peck** and it takes 4 pecks to equal a bushel. So, a bushel is **32 quarts**. It takes 3.28 bushels to equal a barrel, or, I had to haul out about **75 barrels** of squash every other day! It certainly makes me appreciate my job with KTRH!

STRAWBERRIES

I'm putting strawberries in here because everybody seems to love 'em. I can't call the strawberry we eat a fruit or a vegetable. To get technical, the part we eat and call strawberry is actually the swollen tip of the flower stalk. The actual fruit part are the tiny pin-like seed capsules that cover the outside of the berry.

It hasn't been easy to grow strawberries for most beginning gardeners because the strawberry plants are so very temperamental. In this climate, if the sun is too hot or the temperature changes in a heartbeat (as it usually does) the strawberry plant rarely survives.

I'd go to the real experts, the established nurserymen in your area to get the full "skinny" on types of strawberries that will grow in your exact area as well as how to care for them. And that, is the best advice I can give you on growing strawberries.

TOMATOES ···

There is a lot to say about tomatoes because there are so many varieties of tomatoes and they are the favorite among first-time gardeners and can be grown anywhere. In fact, here's a tomato story for you.

My publisher lives in Alvin. When we work on the books I write, each visit is an all-day affair. If I have the day off, I bring my golf clubs and we try to squeeze in 18 holes, then work on the book. If I leave after my weekday program that ends at noon, we play only 9 holes and manage to get four or five hours work done.

When we got to the part on tomatoes, and planned to talk about how easy they are to grow, he shared this tale with me about his experience with the durability of a tomato plants. I believe it.

John, one day I was driving along the Gulf Freeway just before the Dixie Farm Road cutoff coming home, and I got a flat tire.

The traffic was heavy and I was in the speed lane and couldn't pull off the highway so I slowed down and pulled to the side against the divider. I (carefully) got out of my door and waited for someone to stop and help. While waiting, my eye caught sight of some greenery growing between the cracks in the cement. It was a tomato plant with 3 small green tomatoes on it!

Will they grow anywhere? Dirt, sand and grit was the soil, direct sunshine with no shade—ever, and occasional watering (from rain) and the cooling factor of the automobiles as they whizzed by at 60 miles per hour and the darn plant was bearing fruit. I made a feeble attempt to uproot it but decided to leave it alone. I hope someone got to eat the tomatoes when they ripened.

From that story alone, many of you will feel bad when your plant dies. But if you start off with the right location, prepare the soil, keep it watered and fertilize it every week or so, your chances are high that it will bear fruit. That freeway tomato plant was exceptional.

Begin by choosing the right tomato plant for your area. The fact is, there are several good tomatoes for different parts of Texas but for now, let's start off with the Houston area and maybe a 100-mile radius of Houston.

To begin, I'd find a area to plant, say tomato bushes, that was maybe 2' wide and about 10' long. This is sufficient room to plant 5 tomato plants and have enough to feed a family of five. If one or two plants die, or bear lightly, you'll still have enough for a family of five.

Let me tell you, if you want a really terrific vegetable garden you can do it no matter if you've never done one before. The planting itself is easy; just dig a hole in well-prepared soil and put the plant it. It'll grow. But to have juicy, large, well-shaped tomatoes, follow what I'm going to tell you and you won't go wrong.

Types of tomatoes for most areas with the Houston/Galveston climate are: *Spring Giant, Better Boy, Big Set, Floramerica, Freedom, Traveler, Terrific, Homestead 24 and Walter*, for the large tomatoes. I also like the *Beefeater and Beefmaster* that grow to as large as **2 pounds!** These are labeled large fruited specimens.

The main problem with the large fruited tomatoes is that they score (crack) more easily if the summer sun gets hot (like 100"s plus). They still taste about the same if you get to them quickly enough before the sun dries the inside and sucks the juices out. Yes, they're still good but—they look so bad.

For the smaller tomatoes: *Patio, Sweet 100 Hybrid, Small Fry, TAMU Chico 111, Improved Summertime or Hot Set.* They are as tasty and maybe get up to as much as one pound, usually on the plus side of a half-pound is their range.

Then there are the **Cherry tomatoes** that I like to eat off the vine while I'm playing in my garden. There are so many types from which to choose. The cherry tomatoes they sell in local stores are good for this area, so any cherry tomato plants you buy will do well.

But, if you're picky about choosing the right kind of cherry tomato, you can try *Sweet 100, Tiny Tim, Toy Boy, Pixie, Small Fry, and Sugar Lump*, to name a few.

Then, there's the *Big Set, Jackpot, Better Boy, Terrific, and Nematex*. Before we get too far into these tomatoes, let me educate a bit about the letter you might see on the seed packets or on the little stick-in plastic identifier on the tomatoes you'll get in small plastic pots.

I'd advise that you select tomatoes that are medium sized, and the cherry tomatoes. I'd also throw in a few of the *Spring Giants* and *Better Boy* to see how the big ones do. **Be aware** of the letters V, F, N, T, because they mean something. They indicate the **disease resistance** to (V) Verticillum Wilt (makes the leaves curl up), (F) Fusarium Wilt (makes the leaves curl down), (N) Nematodes (microscopic worms in the root system), (T) the Tobacco Mosaic Virus.

Take a few moments and read about preparing the soil for tomatoes. It's the best time you can take to find out before you plant—how to plant!

The two major planting times for tomatoes is Spring, between March 1st and March 30th, later if you have an extended winter or expect a freeze. And for fall harvesting, the last two weeks of August.

In preparing the soil, tomatoes need well-drained soil and a sunny location. A buddy had some terrific soil in a bed right next to his hot tub. The trouble was, the hot tub was under a large oak tree but—sunlight filtered through. As an experiment, he planted some of the hardy tomatoes. He watered them and watched over them and the plants grew!

What I'm saying is that the plants grew, but they *never yielded a single tomato.* Yes, sunlight—direct sunlight is essential. Now, be careful of this Texas sun. Remember, it gets so hot many days that it'll totally burn up your entire garden.

In purchasing tomato plants, choose only the ones that look the strongest. If you start from seeds, start off maybe 5 or 6 weeks before you plan to set them out and make certain is a week or two after the last **frost date!**

> If you have a greenhouse or you have them indoors when starting the seedlings, take them outside a time or two for a few hours each day to "condition" them to the shock of being transplanted. It's called "hardening off".

Tomato seedlings like to be planted deep. Take off the side leaves and bury the plant as much as half of its height into the ground. Space the seedlings about 4' apart, depending on the plant. I like extra room for the plants to bush out and grow tall and not grow "into" another plant.

Tomato rows should be 3-4' from the others. Again, I like room because when the tomato branches start to grow, it's difficult to tell which branch belongs to which bush and it gets confusing. Too, keep a handful of those little plastic coated wire ties to keep the branches from sagging. On a large tomato plant you might use as many as 30 of these ties. They're inexpensive; get a few packets or a long spool of them.

You've heard the term "you little sucker" used in many movies. Well, the "little suckers" on tomato plants are simply beginning offshoots that would form another branch and should be pinched off when they begin to form. You grow bigger and heartier tomatoes if the suckers are gone.

But, leave these "little suckers" on if you need to shade most of your plant, if it's exposed to strong, direct, hot sunshine. It might save the bottom portions of your plant having the almost-ripe tomatoes from *scoring* (cracking).

Watering tomatoes is easy enough; water to where the soil in the root zone is damp but not soggy. I'd water the seedlings every day or so until they gain strength. You should water well-established plants deeply about every week or so.

I know it looks like you're doing a thorough job when you water with a hose or sprinkler system from the top because with the leaves dripping water it seems that "everybody" gets to drink. The fact of the matter is that if you water deeply—to the root system—the water is carried up through the stalk, to the branches and to the leaves. They like it this way too and are less likely to develop a variety of leaf diseases.

In CARING for these tomato plants, here are a few tips. Tomatoes like to climb. If you have say 4-6 plants, as the tomatoes start to grow, you can get by with some little sticks you might find lying around the yard. But as they continue to grow, you may want to look into the taller poles that go up to 6' tall (maybe taller) and I'd opt for those.

These single poles are best if you pinch off the little suckers. Then, your plant will grow tall and these poles are perfect for them. Make certain to use the plastic coated wire ties on the main branch as it climbs and that should do it.

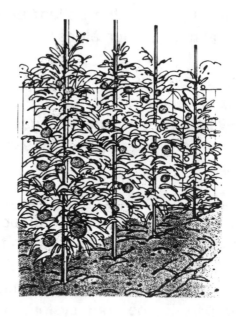

If you use these 6' stakes, put them in at the time you plant or if you've already planted, be careful to put them in about a foot away from the root. If you wait longer, you could take the chance of damaging the root system or killing the entire plant.

If you have access to bamboo cane, you might try to construct something like this. I'd use a galvanized wire to make the ties at the top, and at the bottom, just stick the pole an inch or so deep in the dirt. As the tomato plants grow and you put those ties to the pole and the branches of the bush, it will hold.

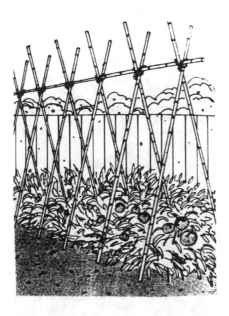

If you have to run your plants along a single row, an idea is to get some of these steel (not too large and painted green; you'll see them at almost any lumber yard or nursery) fence posts and run strands of wire every 18" or so up to the top. Then, you have a type of *fence* to attach these plastic coated wire ties to your plant and these crosswires as your plant grows. Some people use either chicken wire, or what is commonly called **hog** wire to stretch between these steel posts.

Yet another method used to "contain" a tomato plant

to make certain it's wandering branches don't break off or drag the ground when the fruit grows, is to encase it in wire. I like this hog wire because it is large enough to reach through to pick the fruit, yet strong enough to tie your tomatoes with those plastic coated wires. The drawing below is of a homemade support.

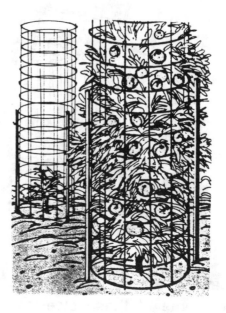

Available also are the ones you can buy that come in various sizes resembling the skeleton of a tepee. They cost about $2 to $4 each and you put them over your young tomato plants, especially the ones where you prefer (or feel it's necessary) **not** to pinch of the suckers therefore giving you a "fatter" plant.

When you see this "tepee-thing", put it over your

young plant from the small side down and stick the ends of the wires into the ground. As your plant grows and you use the ties to secure it to this wire support, it will become sturdy. You'll see these "things" in all garden centers.

In fertilizing your plant, the preparation of your soil is essential and you will not have to fertilize until the tomatoes begin to develop. You will then want to apply a low-nitrogen fertilizer that is high in potassium and phosphorus. A **high-nitrogen** fertilizer will promote lush foliage but will discourage fruit production.

If you want to save yourself some trouble, I'd advise that you get some *Easy Grow Tomato and Vegetable fertilizer* that will feed your garden for the entire season. (**Call 864-7771** for a dealer nearest you).

The temperature will affect tomato production; if it's sunshiny with moderate heat (75-90°F) that's okay. But when the temperature gets over that 100° mark, it hurts all crops and you might want to look into the purchase of some type of shade cloth to cover those vines.

There are a number of problems with growing tomatoes that need to be addressed quickly. There are always worms and bugs that want to eat the tomatoes as much as you do. There's the **Tomato Hornworm** that are disgustingly fat worms, maybe 4" long and the exact color of the tomato foliage.

They have a reddish horn on their hind end and a few faint white stripes. They eat the leaves but also the green and the ripe tomato itself. If you listen closely, you can hear them munching. The surest method is to pluck them off and stomp them to worm heaven—or that other place.

Tomatoes also attract aphids, cutworms, flea beetles, leaf miners, nematodes, whitefly, fusarium wilt and verticillium wilt. I recommend a Pyrethrin. Just a few pages ahead on the section titled **Sprays and Pests**, are some photos of the more common pests and how to handle them.

More information on tomato plants is that there is the **determinate** and the **indeterminate** plant. The determinate usually grows from 3-5'. The **in**determinate plant can go on forever. When it gets to be about 6' tall, cut off the top foot and replant it; you'll have another tomato plant.

The *Big Boy* and *Better Boy* tomato sound great and does yield large fruit (many over 2 pounds per tomato) but they take longer to yield, they tend to score (crack) more in hot Houston summers and, well, they are not my favorite choice.

A tomato grower right here in Houston came up with a truly large, delicious tomato that can take hot sun called the **Sunmaster**. Why not give it a try? A good source for answers to your vegetable problems is **Dan Loep** (Lip) at Covington's Nursery on Airline **(447-1690)**.

Although it seems like a major project as I read over what I've written, it really isn't. A little planning, slight care and watching over your tomatoes can produce a beautiful, bountiful, delicious crop.

Garden centers and nurseries are chock full of information on the various tomatoes they sell and it's best to ask. This book is for home use, for you to read to decide what, when and if you are going to have a garden. It's a ready guide to getting started and to enjoying your vegetable garden. Now, let's see what other delicacy awaits us in the vegetables (maybe fruits) that end in letter after "T" for tomatoes.

Some Texans claim that their tomato plants grow as high as a tree. If so, call Foster Tree Service. Just punch in FOSTERS on your telephone and he'll answer.

TURNIPS ·

Turnips are great vegetables because you can eat the entire plant, root and tops on some varieties. These are *Purple Top White Globe, Just Right, and Tokyo Market*. The turnip to plant for the **leaves only** are: *Seven Top, Crawford, and Shogoin.*

Turnips are excellent for nutrition and have very few calories. They can be easily grown from seed. Plant them in rich soil (pH 5.5—6.8) about ½" deep. Actually, you can "scratch" a small trench (more like an indentation) with a rake handle and cover the seeds by pushing the sides of the trench together with your hand.

As the seedlings appear, thin them out to maybe 6" apart for the **root** turnips and only 4" apart for the **leaf** turnips and rows that are maybe 2' apart if 4" is enough room to get your body into to stoop over and harvest them.

They are ready to harvest anytime between one and two months after planting and yields from 8-12 pounds per 10' row.

Packets of most vegetables contain between 60-150 seeds. It isn't necessary (you understand) to have a **10' row** of anything but either save the seeds and store in cool dry place, top taped shut for next year's planting, or toss them away.

WATERMELON ·

I know this isn't a vegetable but they are delicious and are often grown in vegetable gardens by the home gardener and grow well in this climate. If you have the space for the vines to roam, try them. You can get a 6-pack of watermelon seedlings from most lawn and garden stores or nurseries and if they do grow, you'll be thrilled. That's a promise!

Another friend tried growing watermelons three years ago and from three vines, he got 8 watermelons. They weren't large (maybe the size between a soccer ball and a softball) but they were sweet and he was proud. Here's that watermelon story as told to me by my friend:

A prankster sneaked into my garden one day and put a store-bought watermelon in the midst of the ones I had been bragging about. I looked at it, wrinkled my brow (because I couldn't recall it being that large a day or two before) and approached it with caution.

I counted the melons again and instead of 8, I now had 9. Had I missed my count the day before? What did I use for fertilizer that could prompt such growth in a day or two? As I got closer to the melon—maybe 6' away—I saw a white spot at the bottom that turned out to be the end portion of the grocery sticker. I grinned and let it stay there to pull the trick on my wife. What are friends for?

ZUCCHINI

Zucchini comes in a variety of shapes and colors; baseball-shaped with light green stripes; top-shaped and dark green stripes, and cylindrical and golden. Harvest the ordinary and golden zucchini when they are 4-6" long. Harvest the round zucchini when it's 3-5" in diameter, and the zucchini-scallop squash hybrid when it about 3" across. If it gets too large, it great to brag about and show to people but not as succulent. In fact, it's not good at all!

SPRAYS and PESTS

Now that everything is planted, we have to look for "things" that will destroy these plants. Other than the weather (assuming you prepared your soil and are watering and weeding regularly) the problem now is with bugs. They come in all sizes, shapes and colors. A simple method used successfully by the majority of my listeners who call my program is that most vegetables can be treated with *Malathion* or *Diazinon* spray. On vegetable plants (unlike fruit trees) don't spray to prevent bugs; use these sprays when you see a problem.

If you see bugs on your plants, this doesn't necessarily mean they are harmful; they could very well be beneficial insects that help your plant. The best way to know if these bugs are destructive is:

(1) If you have holes in your leaves.
(2) If your leaves are being chewed off.
(3) If your produce has holes in it.

This is a list of the more common pests that cause damage to vegetables and what you can do to stop the damage. I use the plural (aphids) because I don't ever recall seeing *one* of them on a leaf. They're easy to spot because you'll see an army of them on your leaves.

Aphids are about the size of a the tip of a sharpened pencil, and they come in a variety of colors (pink, black,

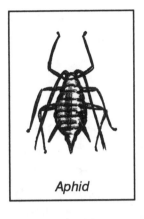

Aphid

green, yellow and various shades in between). They suck the juices from the plant and may transmit viruses. They should be "taken care of" immediately.

The best way to "treat" (get rid of) *Aphids,* is to wash them off your plants with a garden hose. Aphids are easy to spot because they are usually on the back side of the leaf on tender growth. They excrete a sort of sticky "honeydew" on the leaves below it. This nectar is appetizing to the ants (including the stinging fire ant). You can use *Safers Insecticidal Soap,* or *dishwashing liquid.* To be certain they leave, give them a small spurt of Malathion of Diazinon.

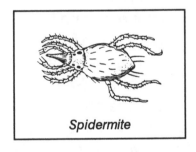

Spidermite

This tiny spider-like creature act like a spider, hence the name Spidermite. They spin webs, suck juices from the leaves and hide under the leaf. Soapy water will do the trick or simple watering might do it but be safe get the good stuff like an insect spray Diazinon or Malathion.

The Corn or Tomato Earworm is simply a caterpillar that is maybe an inch or so long and eats corn, cabbage, peppers, beans and tomatoes (even your roses). They are colored from dark green to black and from light green to

Earworm

pink. They do major damage to the fruit and leaves. Get rid of them immediately!

Offhand, I can't think of a single caterpillar-look-alike that is *beneficial* to vegetables. Everybody recommends Sevin dust and I say the Sevin spray will do the job also. A rule-of-thumb I suggest is when you see crawling bugs you can't identify or if you see leaf or fruit damage, give them a shot with that Malathion or Diazinon spray. For the caterpillars, Sevin dust or spray.

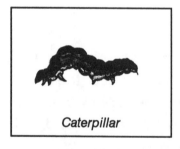

Caterpillar

This worm or caterpillar looks the same as the corn earworm in a black and white photograph but it's color is light green with a few white or pale yellow stripes. Get rid of it too! It eats on (what else?) cabbage, lettuce, tomatoes, beans, cauliflower, broccoli and most other leafy vegetables. For this guy, use Sevin or Diazinon. If you want to get fancy, look for Dipel or Thuricide. For those of you who like an organic garden, Dipel and Thuricide are biological controls. I'd favor the Sevin dust or spray because it will get to the "looper" or the "earworm".

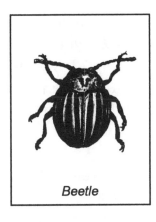

Beetle

The Potato Beetle eats leaves of potatoes, eggplant, bell peppers and tomatoes. They are yellow and black striped and about a large as a dime.

The larvae, smaller of course, are a bright orange color with little black dots and they have a few dozen brothers and sisters, usually on the back side of leaves. They will wipe out a plant in just a few days. Ortho says Sevin dust or Sevin spray will get rid of them and to be certain, spray as often as you see them until they are gone. If you see large ones, you might try using gloves to handpick them or flick them off with your finger. I'd use Diazinon if it's handy.

Leafooted Bug

This guy is a close relative to the Leafooted Plant Bug. Both need to be eliminated. The Bean and Beet Leafhopper is a light green color (some varieties are brown) and they completely annihilate a plant. After they feed and leave their large deposits of honeydew, it results in your plants looking sunburned; the leaves and most of the stem turns yellow to brown and the plant dies. You can use Diazinon or Malathion and spray the tops and bottoms of each leaf.

The Leafooted Bug is a close kin to the Stinkbug and Squash Bug and they do major damage also. If you notice your tomatoes cracking or scoring ("catfacing") the stinkbugs have been there. To make it short, I'd use Sevin dust or spray to see if it does the trick. If not, go to the big gun; Pyrethrin spray.

Stinkbug

This family of stinkbugs really do, in fact, stink. If you squash one, you'll know it. And some look like little armored insects with flat heads, a white stripe over a brown body and if you hit a leaf with one on it, chances are it will fly to the ground. I stomp on them or better yet, use the sprays designed for them and get rid of them forever.

Now, there is one little "stinkbug" that is beneficial called a *Predaceous Stinkbug.* He's dark brown to black with an orange circle around his rear end and with a needle for a nose that resembles a bee stinger. He attacks many bugs bigger than he is and will help. However, many "get the ax" the same as the others with a variety of sprays.

When it's all said and done, there are several sprays and dusts that are especially formulated for these vegetable and plant pests. Malathion and Diazinon will take care of most and Sevin does much of the rest. If you have a little critter that cannot be controlled, put on your gloves and pick one off, put him in a jar and take him to your closest nursery for identification. Then use whatever to kill them.

VARMITS

These entertaining often bothersome little creatures can't be left out because they do wreak havoc to gardens, especially country gardens. You have to make plans for an invasion by raccoons, rabbits, squirrels, armadillos, gophers, and (ugh!) even rats and mice. Although a bird isn't really a varmit, it is however, a pest.

Country gardens is where all of these pests can be found but, for the most part, country gardeners know how to handle these problems. They use live traps such as the Hav-A-Hart model with both ends open; or not-so-friendly traps that kill; they might use a .22 rifle or pellet gun or poison (found in most hardware and feed stores). Although I don't advocate killing animals, country folk oftentimes rely on their gardens to feed their family.

I prefer the use of live traps for country or city gardens because I love animals and even though they are pests, they are just trying to stay alive. Our problem is how to dissuade them from eating our vegetables and either learn to grow their own food or find replacement food not in our gardens.

Birds do major damage to ripe tomatoes. One solution is to hang those old-timey christmas balls on your tomatoes before they ripen. Enough pecks on something that hard will discourage most birds. Some folks put a bird bath near their garden, hoping the birds would drink the water and pass up the tomatoes but the results were varied. One person reports it was like "party time" for the birds. They'd stop in for a drink, peck on the ripe tomatoes then go home and teach the baby birds to do the same thing.

I've heard of good success with those *whirly* things that make noise when the wind blows. Birds don't like noise or movement. And you can get those fake owls or rubber snakes or wind chimes. Some of my listeners report that for a short row of tomatoes they buy a *netting* that both covers the tomatoes to protect them from birds and also to block out some of the intense sun we've talked about.

To enjoy your garden, I'd make a game out of it and see what works best. The rubber snake, fake owl, whirly things, noisemakers and even the bird bath might work for awhile but when the birds get accustomed to it, they still come. Move the gadgets around, take some away, put a new one in and have fun with sort of a contest between you and the birds. If you grow cherry tomatoes, you can afford to lose a few dozen to the birds anyway.

City gardens have many of these pests too. It seems that in almost all areas you'll find raccoons, squirrels, an occasional rabbit, an armadillo or two, a gopher will pop up here and there and, of course, birds.

A friend of mine had some raccoons that got into her attic. She didn't want to poison them and the traps were useless. A co-worker of hers suggested she put a radio in her attic and keep it on all night. They didn't mention any particular station.

*My friend reported more noise and more raccoons. She should have turned on **KTRH**, my station that has news and talk shows. However, the music station she selected encouraged the raccoons to have a party, dancing and snacking on dog food left outside.*

Raccoons love garbage, most dog food, corn on the cob and melons. If KTRH doesn't run them off, call your pest control person. If KTRH does run them off, get your teenager to crawl in the attic and look for an entry hole and board it up.

If you have a city garden (or a country one) and you're trying gardening for the first time and need a solution of any kind, the best advice is to ask your neighbors what works for them, or ask at your local nursery. If you can't get the answers you want from those sources, call your local **County Extension Agent** for an answer to what works best in your area.

As far as rats and mice, there are a variety of traps for them. And if your cat roams outside, you'll not have rats or mice. They'll go to the neighbors yard who doesn't have a "live trap". For squirrels, either a dog or a live trap will do. When you catch these animals, take them a long ways from where you live and drop them off in some large field.

VEGETABLES TO HARVEST MONTH-BY-MONTH

Allow me to give you a month-to-month guide to harvesting vegetables that has proven best in the past 60 years in the Houston area. I'm saying **harvest** remember, not **plant!**

The time to plant these vegetables varies, depending of course on the time it takes these plants to be ready for harvesting. You'll have to look at the vegetables I've listed and then at the time it takes them to mature and plant when you get that figure.

January & February: Broccoli, Brussels Sprouts, Cabbage, Lettuce, Turnip and Mustard Greens, Collards, Kale and Green Onions.

March: Time to harvest the green stuff like Mustard, Turnips, Cabbage, and Green Onions.

April: Snow Peas, Broccoli, Cauliflower.

May: It's potato-picking time for the Irish Potato, Green Beans, Beets, (still Onions), Tomatoes, Carrots and Cucumbers.

June: Tomatoes are in full harvest, Corn, Cucumbers, Summer Squash, Garlic, Carrots and Green Beans too.

July: Now for the fruit we all like; Cantaloupe and Water-melon, along with Corn and Cucumbers.

August, September & October: Pick those Okra, Bell Peppers, Peas and Cucumbers. If you have Tomatoes that survived the summer heat, get those in too.

November: The Tomatoes that were planted in August or September (after the really hot days) are ready for harvest. I hope you planted Lettuce, Winter Squash, Corn, Sweet Potatoes and any Greens you enjoy eating because now they are ready to harvest.

December: The Broccoli, Cauliflower and Brussels Sprouts that were planted months ago are ready to eat, as are Salad Greens, Tomatoes, Turnips and Cabbage.

These Houston summers can be extremely hot! Few vegetables (or humans) can take the intense heat if exposed to it for too long. That's why it's extremely important to plant in plenty enough time so as to avoid the long, hot summer days.

Part 4

HERB GARDENS

I guess I need to start off trying to understand the pronunciation of some words in the english language and this herb garden is a good enough starting point.

It seems you never really know when to pronounce what unless you listen to how the "locals" say it and then you follow suit. There's a wristwatch called *Elgin* (pronounced L-gin, like the drink) and a city just out of Austin, also spelled *Elgin* but pronounced L-gen with a "guh" sound on the "g".

Some words that have an "H" just don't pronounce that "H" and then confuse you by (sometimes) pronouncing the "H". For instance, it's the HUM-ble building but it's "H" UM-ble, Texas. And this herb word is pronounced "Erb" not herb. Why don't they just leave the "H" off when they don't want you to pronounce it?

If you want to grow a (H)erb garden, you can do it and enjoy it whether you live in a house (it isn't Ouse, is it?) in the country, a house in the city, a townhouse, apartment, hi-rise, or trailer home. You can have one at your summer cottage in the woods or at your beach house because they are easy to grow.

The most common (and favorite) herbs are: *Sweet Basil, Chives, Dill, Horseradish, Mint, Rosemary, Sage, Tarragon and Thyme.* Now, let me give you some of what you can flavor with these herbs.

Sweet Basil (several varieties)—*Dark Opal, Purple Ruffles, Minimum, Spicy Globe* and flavors are in *licorice, lemon and cinnamon* among others is used for salads, soups and tomato sauces.

Chives are terrific for almost all soups, salads, baked pork chops, chicken dishes, baked potato, any meat sandwich, hot dogs, hamburger, and also flavor beans and stews.

Dill is mostly for fish, shellfish and pickles and also used for sauces, salads, on vegetables, with cucumbers, and in bread.

Horseradish is for putting on that corned beef, pastrami or cajun roast beef sandwich and for dipping a bite-size cut of prime rib.

Mints (Spearmint, Peppermint), for teas and sauces. Mint is great in coleslaw (crush the leaves), in vegetables, excellent with lamb and veal and it makes your breath fresh.

Rosemary is great for seasoning roast, chicken and meat. Tastes great on pork chops or lamb and is similar to mint. Taste it, and if you like it, grow it and cook with it.

Sage (leaves) are for stuffing, cheese, turkey, chicken or baked fish.

Tarragon to flavor vinegar and for salads, spinach, cauliflower, roast, turkey, chicken and egg dishes.

Thyme (Here's that "H" business again, Thyme is pronounced time) has a clove-like flavor that enhances gumbos, stews and cooked vegetables and makes a stimulating pot of tea. With tea, never use a metal pot.

I can't give you a list of all the herbs but these are the most popular. As far as using them for seasoning, I use them all! You can bring life to an otherwise dull dish by using them as decoration and make that taste come to life. If you happen to really "mess up" something you're cooking in a pot, use **curry** to disguise it and change the taste completely.

If you want some real detailed information on herbs, write to:

The Herb Society of America
300 Massachusetts Ave.
Boston, MA 02115

CONTAINERS AND PLANTER BOXES

For patio herb gardens, use wooden tubs or redwood boxes. There are a wide variety of attractive plastic pots from which to choose and they stay moist longer than clay pots. I like the Mexican-style clay pots though and I've had good success with them.

In setting out these various herbs, just be certain you have gravel or pieces of broken clay pots on the bottom for drainage. A superior grade of potting soil will have all the nutrients necessary for herb growth.

Herbs are either **perennial** (all year long) or **annual** (one year and they die). Some annuals make it an extra season or two. These should be planted separately (the annuals in one pot, perennials in the other).

As far as fertilizing herbs, it really isn't necessary if you start with that good grade of potting soil. Too much fertilizing makes them grow larger and maybe more bushy but it takes away their flavor. Let them do "their stuff" on their own. Just read what I'm telling you about sunlight, shade and water and you'll do fine.

It's said that the best insecticide for herbs is *pyrethrum,* and that it isn't necessary to wash these herbs. If they're outside, I'd give them a run under the faucet a minute or two.

WHAT ARE CHIVES?

I never knew the difference between chives, shallots, scallions, and just plain ol' green onion leaves? Here's what my 2100 plus-page dictionary says. See if you know the difference once I tell you about it.

CHIVES—Of the onion family, with small, slender, hollow leaves used to flavor soups, salads, stews, etc. There are onion and garlic-flavored chives.

SHALLOTS—1. A kind of onion. 2. A vegetable, native of Syria, and resembling the garlic. Shallots divide into a clump (similar to garlic cloves) and are used heavily in French gourmet cooking.

SCALLIONS (Little green onions)—Of the lily family, having an edible bulb, sometimes called "bunching onions" that have either small or no bulbs. These, I chop up and call chives.

QUESTIONS AND ANSWERS

There is no sequence to these questions and answers because I feel that all first-time gardeners need a broad knowledge that apply to growing vegetables. So, here goes . . .

Q.   Tell me the secret that a small gardener can use in storing vegetables.

A.   No! It is confusing, troublesome and time-consuming. My advice is to eat what you grow—as it grows and if there's any left over, give it to a relative, friend of neighbor.

I don't mean to shock you right off the bat with my "no" answer but here's why. One book I read says to (1) Dig a pit in a well drained location. (2) Place a layer of gravel or sand in the bottom, maybe 4" thick. (3) Line the pit with straw and store your vegetables in that pit. Then (4) cover the vegetables with another layer of straw, (5) add 6" of soil and (6) then put in a 3" pipe sunk into the storage space to give it ventilation. Is that (no pun intended) the pits?

We're not Pilgrims or farmers on some frozen Tundra and I say eat the vegetables, put them in the food bin of your refrigerator and if you don't eat them in time (or give them to neighbors) toss them out as fodder for your compost bin! Crass, I know, but realistic.

Q.   What supports do I use for my pole beans?

A.   I'd recommend 3, 10' poles driven into the ground about a foot and a half and form a "tepee". Then, plant one plant at the base of each pole. Who cares if they tangle, they'll still produce if you care for them.

Q.   Can vine crops such as cucumbers, eggplant, melons and squash be grown on supports to save ground space?

A.   Yes! I've heard of apartment dwellers who grow eggplant and cucumbers on their patio by using trellis or along the rails of their upstairs balcony. If you have a yard, put them against a fence and put up a few feet of that hog wire at an angle to support the vines.

Q. Do I need a soaker hose or an irrigation ditch for my vegetables or can I use an overhead sprinkler?

A. I like the soaker hose (or leaky hose) because they get to the roots. I like irrigation ditches to avoid both root rot and to prevent the leaves from staying wet because of disease. If you water early in the morning (7-8 am is okay) overhead, it will give the leaves time to dry out before too long and so the sun won't burn them.

If you water overhead in the evening, it's the same as watering your lawn in the evenings; it promotes fungal diseases of all types. I kid about watering my lawn in the evening because if it's a nice day and I don't have to work, I play golf. The fact is, early-morning watering is by far the best; deep watering reaches the root is even better (for grass, flowers, shrubs and vegetables).

———————————

Q. How can you tell when melons are ripe?

A. The best way is to go get up alongside them and try to lift them up. If the stem breaks easily, it's time. If not, lay it down gently and try again soon.

Q. Tell me about mulch. What is it?

A. First, let me tell you about why we use mulch. Mulch
    conserves heat and blocks out weeds. A good mulch
    is black plastic that comes in rolls and can be bought
    at almost any garden center. Another good mulch is
    newspaper. Some gardeners put the newspapers
    between the rows of their garden to keep out weeds.
    I think it looks messy.

    Other mulches are sawdust, ground corn cobs, pine
    needles (there's plenty of them around and at last
    they're good for something other than killing grass
    and keeping your roof wet), salt marsh hay, peanut
    hulls, seaweed, leaves, and ground bark.

    ───────────────◄►───────────────

Q. Will you give me your definition of a **trench**, **furrow**,
    a **drill**, and a **hill**?. I hear these terms used often in
    gardening books but nobody tells me what they
    are—exactly.

A. **Trench**—a ditch-like excavation 6" or more wide and
    a foot or more deep, usually made with a a plow or
    a good shovel. In the military, a trench is much
    larger, depending on the size of the person hiding in
    it.

Furrow—the hollow between the ridges (hills) of soil that are thrown up in the process of digging. These are the actual walkways between the hills or rows of plants. Oftentimes called a trench.

Drill—the shallow mark in the soil in which seed is sown, usually about an inch deep. I make these drills with a rake handle when setting out seeds.

Hill—usually described as a low, broad, flat mound in which seed is sown. It is a rise in the rows where you plant, higher than the furrow and formed when you dig out the furrow.

Q.   Should I water vegetable seeds after sowing?

A.   Let's define sow in order for everyone to understand what we're talking about. To sow is to scatter seeds on top of soil hoping they make seedlings. To plant, is to put these seedlings in the soil, hoping they'll grow and make big plants.

If you sow (scatter the seed on the soil and lightly cover them) and these seeds grow (into seedlings) you must then thin the seedlings. If you sow these seeds in pots and the seedlings appear, you then plant the seedlings and thinning isn't necessary because you plant a distance apart thereby spacing the seedlings as you plant them.

The answer is that I'd advise against it. Prepare the soil before you sow and make certain it's wet. If you water after you sow, chances are you'll scatter the seed in every direction since they are planted so shallow.

———————————— ‹ › ————————————

Q. You answered my question on a **hill**, but what about the term **hilling?**

A. Hilling is kinda like a short furrow. You should start your squash or melons on the "hill" so the water will drain off.

———————————— ‹ › ————————————

Q. This is a statement and a question. (1) I hate weeding a garden. (2) Are there any weed killers that will not harm my vegetable plants if I get just a little on them?

A. Answer to your statement—everybody does! And to

your question, No.

Q.   I get fire ants in my garden. What is the fastest and surest way to get rid of them?

A.   I can see that you didn't buy my book, YOUR FRONT YARD, because it has a lot about ants in it including the dreaded fire ants.

What works for me and all of my listeners is to use a liquid drench of Diazinon or Malathion. It's fast and thorough. For fire ants that are out of, or near the garden, I love Orthene or Dursban.

Q.   What tools do I need to begin a 10x20' vegetable garden?

A.   The basic iron rake, a hoe and a light, long-handled shovel are all you really need. My picture is on my other book with a flat-tined spading fork and many gardeners say they can't live without them.

If you are gardening for the first time or intend to

have a small garden, I'd shop for tools that are modestly to low priced. You can get a great shovel for less than ten bucks and a hoe and rake for about the same price.

It's fascinating (to me) to walk through these stores and see all the different kinds of tools and contraptions they seem to constantly be coming up with. You don't need them but if it strikes your fancy, get a few extra tools.

One item I'd strongly recommend is either **knee pads** or a **knee cushion** for kneeling in the dirt. I'd also get a pair of garden **gloves.** As you grow more experienced and get a larger garden, you'll want a second hoe, one with a large blade for "hilling" and one with a smaller blade for getting weeds out from between your plants.

You might also buy a second shovel with a flat front for scooping up dirt in the furrows to put back on your hill. And maybe a spading fork for breaking up clumps of dirt and for digging up some potatoes, turnips, etc.

Of course, I didn't mean to leave out a **wheelbarrow** for carting that fertilizer, mulch, pots, etc. There are also different size **gardening carts** available that might be easier for you than using a wheelbarrow. I'd also get a good pair of gardening snips, the kind you can snip off small branches and to cut some vegetables from vines or stalks such as bell pepper and okra.

For spraying insecticides I'd suggest a one-quart for small gardens, and for larger ones, a one-gallon size sprayer that you pump up, hold with one hand and spray with the other.

Q.   How can I get my rows of vegetables straight?

A.   The best method I've found is to tie a string to two stakes and look to see if they are straight as you're digging them. Then, to measure the distance between plants, you can begin with a simple wooden yardstick that you can get for a buck (some stores give them away) and measure. Once your rows are straight, your plants should be nearly as straight.

Whether your garden is straight or not won't affect the yield one bit. We're not pouring a concrete slab, just enjoying a vegetable garden.

Vegetables that grow rapidly taste better and are less prone to insect infestation. In fact, *Medina* makes all sorts of additives that might be right for you. Their manufacturing plant is here in Texas and you can call and they'll send you a catalog of their vast array of fertilizers and soil additives. Their number is 210-426-3011.

You might also call *Ortho*, 1-800-225-2883, for their free catalog of products including sprays, pesticides and herbicides and all the "cides" they have.

For the location nearest you to buy the *Soil Pro (200, 400, 600)*, call 1-800-829-0215.

And, if you want to talk to me during my show, Gardenline, call Monday thru Friday 10am till noon or Saturday from 8-11am at **526-4-740**. If you're long distance try **713-630-5-740** and if your cruising in your car using a mobilephone, it's free if you just punch in **STAR-4-740.**

Maybe you can start a garden with a friend and sort of play a game, you know, compete against each other in a friendly way and see who grows the best vegetables. This way, you and a friend can buy one set of tomato plants that are large and one of the cherry tomatoes and each take half, so you can have variety.

Other than this book, pick up pamphlets at your local lawn and garden store or nursery. The newspapers will give you a good idea on when to plant because the nurseries will begin advertising when it's time or almost time to start.

Closing Message From John Burrow

There really is just so much you can say about vegetable gardening. Various authors will take a different approach and some may have tidbits of information to share with you and "ways" they prefer to do things. Chances are, if they're professionals, they're all correct.

In this book, what I tried to do was make your first experience with growing vegetables reasonably simple and trouble-free. I could have gone into greater detail but I want you to enjoy your garden. The information I've shared with you in this book will get you off to better than a "good" start, I promise you that.

And thank you for indulging some of my down-home stories about when I was a kid. These incidents may not help you, but they were fun for me to remember the days way back when; days, times and people I hope I'll never forget.

I really appreciate those of you who bought this book. Call the program for a question about anything that grows in the ground that wasn't covered in the two books I've written and remind me that you bought one (or both) of my books so I can thank you personally.

I wish you the best of luck in everything you do. As far as vegetables go, do call in and brag on how well your garden is doing.

About the Author

JOHN BURROW was born in Tulia, a small city in Texas, about 50 miles south of Amarillo on a 160 acre rented farm that grew cotton and maize. From talking to him, it doesn't seem that he enjoyed farm life (what kid likes hard work?) but apparently, much of what he learned as a young boy has carried over into his chosen field—talking about the growing and care of plants and flowers. But, as he related stories to me about growing up, spending so much time with his daddy, was special to John.

If you listen to John over the KTRH Gardenline program, you'll learn several things. First, that he knows what he's talking about and second, if he doesn't know, he'll tell you so then find out about it for you. The third thing I feel you'll discover is that John is, in fact, a "country boy" and he's proud of it. And lastly, that John is a "real" person. He is truly a celebrity although he doesn't act like one.

He's come a long way from that rented farm in Tulia to being co-host on the top garden show in all of Texas with far over 2 million listeners who, when they call in, all seem to know him—or at least they "feel" they know him.

His easy-going style, his casual demeanor, his politeness and care to all callers is evident and is, frankly, exactly the way John is in real life. He says he doesn't act like a celebrity because he doesn't feel he is one.

John's first radio show was on KGNC in Amarillo and lasted from 1974 to 1986 where he did a Farm and Ranch show. "Yep, I did a Cows and Plows show and talked about horses, cows and hogs; about hay and feed and manure, and a lot about corn and cotton. I was on from **4am** until **6am** six days a week. It was a living. But when I got a call from KTRH in May of 1986, I jumped at it and have been doing the show ever since.

John Burrow and I have become golfing buddies when he is working on a new book. When I take him to my club, few people recognize him but when he begins to talk, it's surprising how many know him from his voice. John blushes when he's complimented, graciously signs autographs when he's asked, and is very shy.

I talk to John at least a time or two each week to see how he's doing and I ask him to mention his book over his program. "I'm not a very good salesman", he replies, although he feels that each book is good and that it will help many people with their gardening problems. Yet, being shy, he hesitates to "push" it.

Each book reads in the same easy-going manner that John displays over his program. There are many gardening books written but with John's books, they are packed with information and not only act as a ready guide to problems but are also fun to read. His **YOUR FRONT YARD** book has made my front yard much nicer and I'm able to grow some terrific vegetables from following his advice after editing his vegetable book.

If you'll buy a few copies of John's books and give them out as presents, he'll write others and I'll see more of him. I'd like that. You see, other than his editor and publisher, I'm also proud to have him as a friend.

Pete Billac

Other Books by Swan Publishing

HOW NOT TO BE LONELY . . . If you're about to marry, recently divorced or widowed, want to forgive, forget or both, this is an excellent book to read. Candid, positive, entertaining and informative (over 2 million copies sold) $9.95

HOW NOT TO BE LONELY *TONIGHT* . . . Aimed at the *MALE* reader. Other than being courageous and strong, smart women want their man to be sensitive, caring, and understanding. "The" book to give to your man. Or, for men who really want to learn what turns the modern woman on $9.95

NEW FATHER'S BABY GUIDE . . . The "perfect" gift for ALL new fathers. Explains about lamaze classes, burping, feeding and changing the baby plus 40 side-splitting drawings. Most of all, it tells dad how to SPOIL mom! $9.95

YOUR FRONT YARD . . . A fun book of information by KTRH Gardenline co-host John Burrow, tells about plants, trees, grass, pesticides, fertilizer, everything you need to know to win the KTRH Garden of the Month contest $9.95

HOME IMPROVEMENT, *Homeowner's Most Often Asked Questions* . . . A book by home improvement expert Tom Tynan, that answers questions on repair and maintenance of your home. Many readers report that it has saved them hundreds and even thousands of dollars in costly repairs $9.95

BUILDING & REMODELING, *A Homeowner's Guide To Getting Started* . . . Another Tom Tynan book to help you begin a larger project whether it be remodeling a kitchen or bathroom, or building a home from the ground up! Advice on contractors, financing, tips, tricks and mortgages. It **will** save you time and money $9.95

BUYING & SELLING A HOME, *A Homeowner's Guide To Survival* . . . For the buyer and seller; answers you might not be able to get from anyone. All about contracts, agents, inspectors, location, costs, minimum requirements and how to increase the value of your home. A splendid book even realtors found informative and valuable $9.95

IMPOTENCE IS REVERSIBLE—*FOREVER!* . . . by Dr. David Mobley and Dr. Steven Wilson, two of America's foremost authorities on impotence. These urologists **specialize** in this area and have successfully treated thousands. A candid book about the latest evaluations, diagnosis, treatments and surgical procedures on impotence. Complete with photographs and full details on how to reverse impotence-forever! $9.95

KEEP IT UP . . . The second book in a series of male sexual wellness by Dr. Mobley and Dr. Wilson (due in September of 1995) on how to **prevent** impotence. A guide to remaining sexually active and a smart book for men who are in their mid-thirties and above $9.95

Publisher's note: When you order any of these books, you will receive information on the books SWAN has in stock as well as upcoming books.

On all mail-order books, please include $2.90 per copy for shipping and handling costs. The address and telephone numbers for ordering are on the next page.

JOHN BURROW is available for personal appearances, luncheons, banquets, home shows, seminars, etc. He is entertaining and informative. Call (713)388-2547 for cost and availability.

For a copy(s) of *VEGETABLE GARDENING* by John Burrow, send a personal check or money order in the amount of $12.85 ($9.95 plus $2.90 shipping and handling costs) to:

SWAN PUBLISHING
126 Live Oak, Suite 100
Alvin, TX 77511

Please allow 7-10 days for delivery.

To order by Visa, MC, AMEX or Discover, call:
(713) 268-6776 or long distance 1-800-KTRH-740.

LIBRARIES–BOOKSTORES–QUANTITY ORDERS
(713) 388-2547 or FAX (713) 585-3738